CHICKENS

CHICKENS

Derek Hall

CHARTWELL
BOOKS, INC.

Reprinted in 2010 by
CHARTWELL BOOKS, INC.
A division of BOOK SALES, INC.
276 Fifth Avenue
Suite 206
New York
NY 10001
USA

**Copyright © 2009 Regency
House Publishing Limited**
The Red House
84 High Street
Buntingford
Hertfordshire
SG9 9AJ
UK

For all editorial enquiries, please contact
Regency House Publishing at
www.regencyhousepublishing.com

ISBN-13: 978-0-7858-2407-7

ISBN-10: 0-7858-2407-3

Printed in China

Reprinted in 2010

CONTENTS

INTRODUCTION

Chickens and other poultry have been kept and reared in Europe for over 2,000 years, and have also long appeared in places as far apart as India and China. Today, there is hardly a place in the world that does not have thriving populations of domestic chickens, and it is easy to see why this should be so. Many chicken breeds are hardy and will also forage for themselves if necessary. They are quick to grow, mature, and feather up, and quickly come to a stage when they are capable of laying eggs or are ready for the table. Many chickens are capable of laying several hundred eggs in a year – with more than 300 in some breeds – and a good-quality chicken will provide plenty of succulent, lean meat that is particularly appreciated in these health-

Domestic chickens have long played an important part in the lives of people all over the world, in that their eggs and meat provide valuable and relatively cheap sources of food.

conscious times. A chicken's egg provides fat, carbohydrate, protein, minerals and vitamins, and as well as being delicious eaten 'whole' is, of course, an important ingredient in many recipes. Likewise, the meat of a chicken is the most important constituent in a huge range of dishes, from casseroles, curries and stews to internationally renowned specialities like *paella* and *coq au vin*. Chicken soup, always a favourite, is regarded as a 'medicine' for those recovering from illness, due to the fact that it can be accepted by the body when other foods might be more difficult to consume or digest. It is such versatility that has made the chicken one of the most important of all food sources worldwide.

CHICKENS

It was not so very long ago that a roast chicken was considered a luxury to be enjoyed mainly by the wealthy or on high days and holidays as a special treat; but nowadays it has come to be regarded as a cheap, everyday food for all. It is not surprising, therefore, that a vast industry has developed to supply the insatiable demand for both chickens and their eggs, which has often been accused of putting profits and productivity before the welfare and well-being of the birds. Among the arguments put forward by the industry in its defence are that the regimented techniques it employs are a necessary part of providing the consumer with an economical source of food, which in any case conform to the regulations set out by those whose role it is to monitor the industry. However, it is not within the scope of this book to delve into the world of the commercial producers, and their methods can be argued from both sides. If we continue to demand cheap food, then we may have to accept that large-scale production has been designed to give us what we want, even though it may very

Free-range chickens live a more natural, happier life and produce better quality meat and eggs than those reared in battery farms.

INTRODUCTION

well offend our finer sensibilities. On the other hand, the produce from chickens reared in kinder conditions is not exactly off the scale in terms of price and is being bought in increasing quantities by a more discerning consumer, who may be rejecting the 'battery hen' in favour of a tastier bird that has had a happier life.

Of course, not all chicken rearing is carried out under the so-called battery system used by the industry as a whole. Many farmers allow their chickens a degree of freedom and a lifestyle that is very different from the more intensively reared, and chickens and eggs produced in such 'free-range' conditions are much sought after by those who can afford to pay the extra price they take to produce. Likewise, people in many countries keep a few chickens for their eggs and meat and simply allow them to scratch around on a free-range basis close to their dwellings, or keep them in enclosures in which they are free to scratch around. Providing they are properly fed, a half-dozen or so chickens can keep a family in eggs while providing a ready source of garden fertilizer at the same time.

Free-range chickens require a safe, fox-proof place in which to roost and sleep at night.

INTRODUCTION

exciting world is revealed, while those deciding on a chicken or two will find many breeds from which to choose. Not all chickens make good pets or are easy to keep: some are fragile or require special care and conditions, while others are downright unsociable and may even deliver a sharp peck if approached too near. But there are many breeds from which it is possible to chose a perfectly friendly and easy-to-keep bird. A reputable breeder will usually advise the kind best suited to your needs and will suggest the type of conditions in which they should be kept.

Many of the world's chicken breeds are described in the following pages, together with their origins, and useful advice is offered on providing for their needs. But the more research you can do for yourself, by talking to experts, visiting exhibitions and poultry shows, and by reading many of the excellent books available on all aspects of the subject, the more satisfying the whole experience will prove to be.

For many more, however, keeping chickens is simply a rewarding hobby, and some positively enjoy having pet chickens about, and for whom the regular supply of fresh eggs is a pleasant bonus. For such people it is also a natural way of introducing children to animals and their care, and even to teaching them some of the inescapable facts of life. Some owners may go on to exhibit their chickens in shows, and those interested in this activity may well discover that a new and

Some like to show their chickens, while others keep them mostly for the pleasure of seeing them pecking around in their backyards.

CHAPTER ONE
FACTS ABOUT CHICKENS

All of today's domestic chickens are believed to be descended from the Asian jungle fowl, of which four distinct types exist in the wild. It is possible the chicken could have derived from any one (or more) of these, although the candidate most strongly argued is the red jungle fowl (*Gallus gallus*), and one only has to look at this to spot its resemblance to many of our modern domestic breeds, a similarity that is most strongly marked in cockerels or roosters (male birds).

Recent research, however, suggests that the yellow skin colour of the domestic chicken may have derived from the grey jungle fowl (*Gallus sonneratii*), strengthening the argument for some sort of dual ancestry. The problem with this proposal, however, is that the offspring of two different species are invariably sterile; was it possible, therefore, that this new strain was somehow able to overcome this biological hurdle and go on to become a new, sexually reproducing species? Clearly, there is more work to be done on the subject by geneticists and other biologists before the issue can be fully resolved. In the meantime, we shall use the red jungle fowl as our comparison with today's domestic chicken, since it exemplifies a lifestyle that is in general very similar in both.

Fowls occurred and were domesticated in various places throughout Asia, including India, where cockfighting was prevalent, eventually spreading throughout the Near East, Africa, and the Greco-Roman world.

food intended for livestock can be readily gleaned. Such forms of bold, opportunist behaviour often mirror the lifestyle and feeding habits of modern, free-ranging domestic chickens.

Like the typical chicken, the red jungle fowl is a robust, deep-bodied bird with powerful clawed feet and and a shortish, slightly down-curved beak. The clawed feet may be used in territorial fights by the males to strike at their opponents, but in both sexes they are well-adapted tools used for scratching and scraping at hard, sunbaked soil to unearth food. A great variety of insects, spiders, grubs, seeds and shoots form the bulk of the diet, but the red jungle fowl is also adept at catching small reptiles such as lizards. The jungle fowl is essentially a ground-feeding bird, and the beak is ideally shaped for selecting and picking up food items lying on the ground. The purposeful way in which it struts over the ground, pausing with head cocked from time to time, will be familiar to anyone who has ever watched chickens feeding.

Again, like the chicken, the top of the red jungle fowl's head bears a fleshy red comb, and two fleshy wattles hang beneath the chin, structures that are larger in the male birds. As befits their need to remain concealed while sitting on the

The wild red jungle fowl is a common and locally widespread bird that ranges throughout much of Asia, from Nepal through India, then eastward and southward to China, Vietnam and Indonesia. It ranges almost entirely over the tropical and subtropical regions of the world, its preferred habitat being low-lying wooded or bushy terrain interspersed with clearings. These environments provide both shelter from predators and a shady respite during the heat of the day, with more open ground on which the bird can forage during the cooler mornings and evenings. The birds may become more tame with local people in parts of the range where they are not persecuted, and are often found around the edges of farms and villages, where items such as the seeds of arable crops or

nest, hens (females) are subtly patterned with brownish and chestnut plumage on the head and neck, but with darker streaks and spots on the body and a black tail, which is held erect in characteristic fashion. Males have a long shawl of feathers extending from the head to about halfway down the body, which may be red-orange or yellowish depending on the race, with red, brown and copper feathering on the body and a long, glossy green and arching tail. The male jungle fowl grows to about 29 inches (74cm), the female, with her much smaller tail, being about 18-in (46-cm) long.

In red jungle fowls the female makes a chicken-like clucking call, while that of the male evokes the dawn crowing of a domestic farmyard cockerel. This is used as a territorial warning both at daybreak, at night and even throughout the day as the bird moves about its territory. The male also calls to exert dominance over other males and to attract females at mating time. Unlike some breeds of domestic chicken, which scarcely become

The Red Dorking is the oldest example of the race of five-toed fowl first described by the Romans. It has inherited many of the typical attributes of the Asian red jungle fowl.

airborne, even when launching themselves from high perches, red jungle fowls are quite capable fliers. Although they prefer to run on their long, powerful hind legs, they will take to the air with a clatter of wings if danger threatens, seeking refuge in a tree and often using it as a place in which to roost.

CHICKEN BIOLOGY

The domestic chicken (*Gallus gallus domesticus*), as we have seen, is closely related to the jungle fowl. Not only are there strong physical similarities between the two, but their habit and way of life are remarkably similar. But let's start at the beginning. The chicken is a warm-blooded, egg-laying vertebrate animal ('vertebrate' means that it has a backbone), and it is placed by zoologists in the class Aves along with all the other types of birds. (The other classes of vertebrates are the fish, the amphibians, the reptiles and the mammals.) Within the class Aves the chicken is placed in the order Galliformes (the gamebirds), which encompasses over 180 different species, including many familiar ones such as the pheasants, quails, grouse, turkeys, partridges, guinea fowl and bobwhites. Different representatives of the order Galliformes are distributed over most of

the world, including Europe, Asia, America and Australia. Within this order, the chicken is further grouped into the biggest family, the Phasianidae, which includes the pheasants.

The chicken is a fairly unremarkable bird by some avian standards, and in that respect could be said to be fairly typical of gamebirds as a whole. The chicken neither flies spectacularly, nor does it have a fascinating or dramatic way of catching its food; it has no ability to

The pheasant is a member of the avian order Galliformes, to which the chicken, together with other gamebirds, also belongs.

mimic, it can't swim, its mating displays only seem to impress other chickens, and it doesn't look especially streamlined or eye-catching. Partly, of course, this 'ordinariness' is a result of the intensive breeding that the chicken has undergone since earliest times to produce certain

FACTS ABOUT CHICKENS

chicken is very much a bird, however, and as such demonstrates all the basic features of the class Aves. Furthermore, although it may seem unspectacular, it is nevertheless perfectly suited to its way of life, in the same way that the jungle fowl has been created through the long process of evolution into a creature superbly adapted for its particular lifestyle.

Domestic chickens are not capable of long-distance flight, although lighter birds are generally able to fly short distances, such as over fences or into trees, where they would naturally roost.

The skeleton
The chicken skeleton, like that of all birds, is modified in several significant

desired traits, making it a uniform commodity for the food market or to conform to standards for exhibition. Even the wild jungle fowl is in many ways an unspectacular species compared with many other birds, although it does well enough in its own way. Anatomically, the

wishbone – a feature probably familiar to anyone who has eaten a chicken or a turkey – and the breastbone itself forms a deep, flattened keel, or carina, providing a broad area of anchorage for the strong pectoral muscles. It is these pectorals that form the main muscles that power the wings in flight, and it is these that provide much of the white 'breast' meat on a chicken or turkey carcass. The vertebrae of the backbone extend forward to form a long, flexible neck, and behind the pelvis the vertebrae of the backbone extend to form a short tail.

The bones of the wings – equivalent to the arm and hand bones of a human being – are elongated, providing the anchorage for the flight feathers. Usually, the feathers on the inner part of the wing (the part nearest the bird's body) are supported by the ulna and radius bones, and the feathers on the outer section of the wing are supported by the bones that are equivalent to the hand bones of the human skeleton, the point at which the

ABOVE LEFT: These two Poland cockerels are squaring up for a fight.

PAGES 28 & 29: Gamecocks, such as these European birds, were originally used for cockfighting.

and different ways from that of other vertebrates, mostly as a means of assisting the flying process, even though chickens aren't renowned for their aerial skills. However, these modifications for flight came about long ago during the long process of bird evolution, and it isn't surprising that chickens retain these features, since they are also present in other types of birds which spend all of their lives on the ground. The bones of a bird are hollow to help save weight and therefore energy when getting airborne and once it is in flight. To offset this potential weakness, the bones are supported internally by a system of bony struts that give them extra strength – not unlike cross-struts that give strength and rigidity to girders and similar structures. The collarbones form a V-shaped

CHICKENS

Chickens may not be great at flying but they are strong and capable runners.

bird's wing bends being equivalent to the that of the human wrist. In the bird's leg, the tibia and fibula correspond to our shin, and below this the joint is the equivalent of our ankle – which is why it hinges forward in a bird and not backward. The bird's tarsus is equivalent to our foot, while the bird's foot is composed of the bones that form the toes in human beings. Most birds have four toes; in the chicken, as in many other bird species, three toes point forward and one backward. Some breeds of chicken, such as the Dorking, Faverolles and Sultan, have five toes, however, although the fifth does not touch the ground. As befits birds that spend much of their life on the ground, chickens are good runners, being capable of speeds of about 9mph (14.5km/h). The musculature needed for running also creates the brown meat we find in abundance on a chicken's leg and thigh. The lower legs of a chicken, like those of other birds, are covered in tiny overlapping scales.

The skull of the chicken has also been modified for lightness: instead of having heavy, bony jaws lined with teeth, the chicken and all other birds have beaks or

FACTS ABOUT CHICKENS

bills instead, which are formed from the bony upper and lower mandible bones of the skull. To give it strength, the beak is encased in a hard, horny sheath, which complements the other weight-saving modifications seen in the remainder of the bones of the skeleton. The beak itself is a remarkably variable structure in birds, designed primarily for catching or picking up food but also sometimes used in courtship, fighting and other activities, such as nest-building and excavating. In chickens the beak is shortish and pointed, reflecting the bird's feeding habit of pecking at the ground for seeds and other food items. There are large sockets on either side of the skull that accommodate the eyes.

Internal organs and physiology

Chickens, like all other birds, do not chew their food since they have no teeth. Instead, the food is swallowed whole, travels down the gullet, or oesophagus, and enters the sac-like structure known as the crop, which is situated at the base of the neck and is where food is stored. From here, food enters the gizzard, a

muscular organ which grinds it up with the aid of secreted digestive juices. Grit is also taken in and swallowed by the chicken to facilitate the grinding process. The food passes from the gizzard into the intestine where digestion and absorption of the nutrients takes place, following which waste matter is voided at the anus in the form of faeces.

Unlike other vertebrates, birds do not have large bladders in which to collect urine, since this would add extra weight to the body and impair flight. Instead, their kidneys are specially designed to produce a highly concentrated nitrogenous compound called uric acid, which leaves the body via the anus together with the faeces.

The act of flying entails enormous amounts of energy being expended and requires the body to work at a high temperature for peak efficiency. Because the flight muscles need lots of oxygen, the lungs are especially well-endowed with capillaries. Furthermore, the lungs are joined to a series of elongated sacs that extend into the bird's internal body cavities, a feature that enables the lungs to hold a greater amount of air, improves oxygen absorption into the bloodstream, and also makes the bird lighter. Most birds have nine of these air sacs.

Feathers and flight

Only birds have feathers, and it is perhaps this single feature that distinguishes them from all other animals. It is also the feature that gives the birds their ability to fly – although this remarkable feat is seen in one form or another in several other kinds of animals. Feathers are formed from a substance known as keratin, which is the same material from which the scales of reptiles and our own hair and fingernails are composed. A bird has several different types of feathers. Flight feathers are large and strong and found on the wings and tail. They are designed to produce a smooth flight surface over which air can flow to enable the bird to

In general, the feathers of birds are quite complex, coming in several different types that perform particular functions.

FACTS ABOUT CHICKENS

achieve lift and also to help propel it forward as it beats its wings. A second type, known as contour feathers, gives the bird its streamlined shape, allowing it to pass through the air efficiently. A third type, the down feathers, provides insulation to help maintain the high internal temperatures needed for flight. The last type, the filoplumes, are found especially around the eyes and beak and have a sensory function.

A typical bird feather has a stiff, central hollow shaft which grows from a follicle in the skin. Arising from each side

BELOW LEFT: These chickens are having a dust bath, which helps to rid their feathers of parasites.

OPPOSITE: This rooster has a well-preened, glossy set of feathers, that also indicates he is in good health.

of the shaft are structures called barbs, and these are fringed with tiny hooked barbules that link together to form the main, flattish section of the feather known as the vane. Large birds, such as swans, have up to 25,000 feathers on their bodies, but other species may have considerably fewer.

For a bird to fly, it must first get air flowing over its wings (the flight surface) at sufficient speed for lift to be created. It does this either by running along the ground or skimming the surface of water, pushing upward or dropping from a perch to create an initial airflow. Once the wings begin to provide lift the bird becomes airborne, propelling itself forward by the down-beating action of the outer part of the wing, while the inner part maintains lift. The tail helps in steering and is also used as an airbrake, when coming in to land, by fanning downward to create wind resistance to slow the bird down. Some birds have developed the art of flying to an almost incredible degree, and birds such as the swift (Apus species) hardly ever land, but eat, sleep and even mate in the air. Others, like the peregrine (*Falco peregrinus*), can stoop at breathtaking speeds to catch prey, while hummingbirds are able to hover motionless in the air and even fly backward. The chicken is no such master of the air, however, although some breeds are able to emulate their jungle fowl cousins and make it onto the lower branches of the nearest tree or even clear a fence in a bid for freedom. Nevertheless, the chicken will never be admired for its powers of flight. Chickens reared in open-air runs or pens may have one of their wings clipped to help prevent them from escaping; this entails clipping the ends of the flight feathers to unbalance the bird so that it cannot remain in the air for very long.

Feathers are absolutely vital to a bird's existence, and they are replaced once or twice a year in the process known as moulting. This ensures the feathers are always in the best of condition, and also that worn ones are replaced. In between times, birds spend much time preening, which includes lubricating the feathers with oil from a gland near the base of the tail, cleaning the feathers to remove dirt

and parasites, pulling them back into shape with the beak, and dust-bathing. Even domestic chickens undertake this regime of personal hygiene, for even if flight is not the main issue, clean feathers mean fewer parasites and more efficient body insulation for the bird concerned.

As already mentioned, the feathers on a bird do more than provide a means of getting it into the air. Depending on the species, they also act as camouflage, assist in display and courtship, and may be used to line the nest to make it warm for incubating eggs and for chicks when they have hatched. Because chickens are bred under controlled or artificial conditions their feather colouring can be somewhat artificial, too, although even in this situation the hens often tend to be duller than males – as would befit a wild bird spending a long time concealed on a nest. Cockerels tend to be more flamboyant, having bolder colours and longer tails, while they also tend to exhibit other male/female differences seen in wild stock; thus males often grow bigger and have bigger or more elaborate

Contrary to popular belief, chickens may not be as stupid as was once thought, and many claim that their own birds display plenty of character and intelligence.

combs and wattles. Today, many breeds of chicken are available in a variety of different colours or variations of colour combinations, sometimes because they

have been specially bred for showing. Another reason is that certain feather colours or patterns are linked genetically with other desirable traits, such as meat

quality, fertility, the ability to lay eggs of a specific colour, or skin colour.

The brain and senses

The word birdbrain is often used to describe someone of limited intelligence, and while some birds, such as parrots and members of the crow family, are considered to be relatively 'intelligent', and are certainly resourceful and capable of undertaking simple tasks, the chicken, and indeed gamebirds in general, are not usually included in this elite avian group. Stories about chickens becoming hypnotized merely by having their heads held near to the ground while a line is drawn in front of them with a stick, or by having their heads tucked under their

Like all birds, chickens have much better eyesight than human beings.

wing so that they will go to sleep, have done nothing to alter the belief that we are looking at fairly simple creatures that are less than bright. There is no doubt that part of this perception arises because few have bothered to examine how smart chickens might actually be, and also because it suits us to regard expendable animals as being rather dim and unaware of their ultimate fate, rather than being highly sentient creatures like some other animals and indeed ourselves. But those who keep and study their pet chickens report many examples of amusing, unusual and 'clever' behaviour, and offer convincing evidence of the individuality and personality that is displayed by their own birds.

For all that, chickens at least share with other birds a sense of sight that is far greater than our own ability to view objects, and we are indeed quite feeble at this task compared with the average bird. The optic lobes of a bird's brain – the parts concerned with vision – are relatively much bigger than those of a human being, and most birds can see objects 30 or more times better than us, these being not necessarily larger but in much greater detail. This helps the chicken to pick out small insects or tiny seeds from similar-coloured stones and

other debris on the ground, and spot movement that might indicate a potential predator. This may be linked to birds' ability to perceive different wavelengths of light from those that we can see. Research with chickens also suggests they have a sense of direction based on the use of the Earth's magnetic fields, an ability that may be present in other birds, too. Birds also have acute hearing,

although in many species the sense of smell is not especially well-developed.

Other structures

There are various prominent features seen in adult chickens, some of which are also present in other birds. First there is the comb, which is a fleshy, usually red or

LEFT: This Hamburg chicken has a rose comb, which is solid, broad and nearly flat on top.

ABOVE: The strawberry comb is set low and well forward on the head.

CHICKENS

The single comb is the most commonly seen, being always upright and much larger and thicker in the male.

purple crest on the top of the head, this often being a rather large and prominent structure in males. The comb is also present in hens, but is usually less elaborate. In males, the comb probably has some sexual, territorial or hierarchical function, but in both sexes it is used to help the bird lose heat from its body. The comb is richly supplied with blood vessels and the blood is cooled as it passes through the comb. Heat is also lost from the chicken's body via the comb, which thus works as a kind of 'radiator'. Combs are delicate structures and of necessity have no protective feathers; in some breeds they can be prone to frostbite in winter.

The comb may be formed into a number of points, or perhaps be V-shaped, the comb varying according to the breed of chicken. The **rose comb** is broad, fleshy and solid, being almost flat on top. Depending on the breed, the comb may turn upward, appear almost horizontal, or trace the contour of the head. The **strawberry comb** is low in profile and set well forward on the head, and resembles the outer surface of a strawberry. The **silkis comb** is round and lumpy, being wider than it is long, and is covered with indentations and corrugations. The **single comb** is a thin, fleshy structure with a smooth and soft surface, starting at the beak and stretching along the top of the head. The top of the comb has a series of distinct points and corresponding depressions, usually five or six in number, giving it the characteristic appearance often associated with a 'typical' chicken – especially a cockerel or rooster. It is held upright in males, while in females it may be upright or fall to one side according to the breed. The **cushion comb** is low and fairly small with an almost straight front, sides and rear and no spikes. The **buttercup comb** has a cup-shaped crown set in the centre of the skull, with prominent points. The **pea comb** is a low comb of medium length, whose top is characterized by three ridges running lengthwise; the centre ridge is slightly higher than the ones at the side, and the outer ridges are either undulated or have small serrations. Finally, the **V-shaped comb** is composed of two horn-like parts joined at their bases.

This Sussex cockerel has a prominent single comb and pendulous wattles.

CHICKENS

There are also flaps of red skin that hang down from the sides of the beak. These are called wattles, some of which are quite long and pendulous. Again the size and shape of the wattles vary according to the sex of the chicken and the breed. Like the comb, the function of

Chickens tend to favour a particular place where they habitually lay their eggs.

the wattles is to help the body lose excess heat in hot weather.

A chicken's head is not entirely covered with feathers: there are bare patches of skin on the sides of the head, and on each side there is a specific area called the ear lobe. The ear lobes may be red or white, and they are a useful feature that help to distinguish certain breeds. As a general rule, chicken breeds that have white ear lobes lay white eggs, whereas

breeds with dark (red) ear lobes tend to lay brown eggs – although this is not a hard and fast rule. For example, the breed of chicken known as the Silkie usually has dark ear lobes although the hens usually lay white eggs.

Reproduction and egg-laying

Chickens usually become sexually mature at about 18 to 24 weeks of age. Free-ranging chickens may show some of the

FACTS ABOUT CHICKENS

A broody hen incubating her eggs, some of which have already hatched or are in the process of doing so.

forms of mating behaviour seen in their wild jungle fowl cousins: the cock may first call hens to him by finding food items for them, then he may drop one wing, fan his tail, and lean and twist over the female before mating with her and fertilizing her eggs. The eggshell, being a tough, waterproof protective layer made from a type of calcium carbonate called calcite, encompasses the egg, protecting it as it travels down the lower part of the oviduct.

Once the hen lays her eggs – in natural conditions usually in a mere scrape on the ground – she may become broody, making her sit resolutely on her eggs in order to incubate them. She will resist attempts to remove her from her nest, and may even be reluctant to leave it to feed or preen herself. The hen will turn the eggs regularly at this time to ensure they are kept at a constant temperature. Cochins and Cornish hens are among the broody types, but many modern breeds do not, or abandon the activity part-way through the incubation period. Domestic chickens which are less prone to broodiness are favoured by egg-

producing farmers, since the hen is ready to lay eggs again more quickly than the broody type.

Chickens become accustomed to laying their eggs in the same place each time. Farmers and other chicken-keepers

LEFT: A chicken has laid an egg within the protection of an old wheelbarrow.

ABOVE: The chick breaks through the eggshell using its egg tooth.

OPPOSITE: The pretty chicks are born covered in a light, fluffy down.

CHICKENS

often put artificial eggs made from stone or other materials in places where they want the chickens to lay. Flocks then tend to use only a few preferred locations instead of each individual bird choosing a different spot. Hens may even try to choose the same nest for their eggs. These are all beneficial behavioural traits as far as the farmer is concerned, since it means that eggs can be gathered quickly and safely without having to hunt around for them.

Fluffy little chicks have a universal appeal, and are fiercely guarded by the mother hen, which broods them when necessary to keep them warm, and often returns to the nest at night when her chicks are very young.

During the incubation period the developing chick gets nourishment from the yolk within the egg. Incubation lasts for about 21 days – about the same length of time as in the jungle fowl – and assuming the eggs were fertilized by the cockerel, they are then ready to hatch. The eggs are not always laid at the same time, so there is a gap between hatchings during which time the broody hen stays on the nest. When the chicks are ready to hatch they emit cheeping sounds to which the hen responds by gently clucking, encouraging them to cut their way out of their shells using their egg teeth. These are small, sharp, raised parts of the upper bill, which become absorbed after hatching.

BEHAVIOUR

As already seen, a free-range or unconfined chicken shares many traits with the jungle fowl. Chickens are gregarious by nature and will form flocks, often of more than one breed. This communal approach to life also extends to

RIGHT: Mother hens call their chicks to places where food is present.

PAGES 52–53: Cockerels aren't always popular, in that they are inclined to crow at any time of the night or day.

FACTS ABOUT CHICKENS

the incubation of the eggs and care of the young. A hierarchical dominance exists in flocks, and individual chickens hold sway over others, establishing a pecking order such that the most dominant birds get priority when it comes to accessing food and the best places in which to nest.

Cockerels are top of the pecking order, and will attempt to surround themselves with a harem of hens, which they will defend against other males. If dominant cockerels or hens are removed from the flock, a new social pecking order will become established in time.

Cockerels are associated with crowing, and the shrill, persistant and penetrating call can be a source of great irritation; in fact, the decision by a neighbour to 'keep chickens' is nearly always viewed with dismay. It is one thing to experience the sound if you have always been accustomed to it, but others find it hard to endure being habitually woken at first light, or even earlier. As mentioned previously, cockerels also crow at night and even during the day, the reason for the call being similar to that which causes

OPPOSITE: Anatomy of the chicken.

RIGHT: Cockerels play no part in the rearing of the chicks.

FACTS ABOUT CHICKENS

Chickens are omnivores, and when living in the wild will scratch at the soil in search of seeds, insects, and even larger animals such as lizards and young mice.

other birds to make sounds: the cockerel is signalling his territorial claims to other males, even though there may be none in the vicinity, while at other times the call is provoked when the bird has been disturbed and is sounding the alarm. Hens, by comparison, are reasonably quiet, gently clucking to their chicks. They may become more vociferous after laying eggs, however, but this is mild compared with the strident call of the male.

Parental care consists of the hen occasionally feeding the chicks, although usually she simply leads them to food and water or calls them to edible items she has found. For a while, she will also protect the chicks, taking them under her wing, but will begin to lose interest in them after a few weeks when they will be left to fend for themselves. At this point, the hen may lay again, the cock having played no part in the rearing of the young. The lifespan of a chicken, under natural conditions, is up to about seven years, although some have been known to live for ten or 15 years or more.

CHAPTER TWO
DOMESTICATION OF THE CHICKEN

Just as the exact ancestral lineage of the chicken must to some extent be informed conjecture, so too must be the process by which an animal like the jungle fowl became domesticated. However, it is relatively easy to imagine the scenario in which this might have taken place. Once human beings began to turn from a purely nomadic existence to a settled, farming way of life, they would have felled trees to plant crops, creating clearings in the Asian jungles. We have already seen how such a habitat is favoured by jungle fowl, and this, coupled with easy pickings in the form of scattered seeds and possibly scraps of food discarded by humans, would have encouraged the birds to frequent human habitation. Our ancestors, in turn, would no doubt have looked upon

RIGHT & OPPOSITE: These are proud pedigree cockerels, and they have come a long way, in terms of breeding techniques, from their ancestor the Asian jungle fowl.

DOMESTICATION OF THE CHICKEN

the jungle fowl as an easily kept, readily available and nutritious source of food, trapping and caging some of them to provide them with eggs or for when they would be required as meat. The jungle fowl, therefore, would have been a most welcome visitor.

It is also just as likely that human beings were already well aware of the eating qualities of the jungle fowl, since the birds were probably already an important part of the hunter's fare, along with other jungle animals.

Human beings were not slow in discovering the full potential of the jungle fowl, the next logical step being to turn it into the useful, domestic chicken that we know and value today.

It didn't take long for humans to realize that jungle fowl were also ready breeders, with chicks that developed quickly, which swiftly led to exploiting them for food. Just as the dog was selectively bred by humans so that desired

DOMESTICATION OF THE CHICKEN

A 5th-century mosaic from Ouza'i, Lebanon, a southern suburb of Beirut, showing domestic animals that includes chickens.

traits would be continued from generation to generation, so it is likely that jungle fowl were also selected for such features, in time producing birds with plump flesh and good egg-laying qualities – in other words, a form of dual-purpose domestic chicken. Despite the chicken's undoubted attributes as a food source, there is evidence that domestic fowl were also kept for religious rituals, and it is well documented that ancient cultures, such as those of the early Greeks and Romans, offered chickens as sacrifices to the gods.

Domestication probably took place about 6,000 years ago: we know that the domestic fowl was present in India by about 3200 BC and in Egypt and Crete by 1500 BC, reaching China by about 1400 BC. There is also evidence that the chicken was domesticated at an early stage in places such as South-East Asia and by the Neolithic cultures of Oceania, and that Polynesian traders brought chickens to Easter Island in the South Pacific in about 1200 BC. The birds then reached Asia and Indonesia as cargo, and were finally introduced to Europe and western Asia.

CHICKENS

An attractive poster, dating from c.1868, that depicts the valuable poultry breeds popular at that time.

By the time of the Romans, chickens were kept in many parts of Europe, where they were used for cockfighting as well as for their meat. Chickens were also used by the Romans as oracles or as mediums through which advice or prophecy could be transmitted to them by the gods. When a 'sign' was required, if the chickens flew out from their cage or beat their wings instead of eating the special food they were offered, this could be interpreted as a bad omen. One Roman general threw his chickens overboard because they refused to eat prior to a battle; but the battle was lost anyway, resulting in a heavy loss of ships.

The Romans even produced books on chicken-keeping, in one of which the writer, Columella, describes several breeds, with accounts of their general nature and appearance, the best number of birds to keep in a flock, advice on the best siting of coops, and much information concerning feeding and general care.

After the fall of the Roman Empire and the advent of the Dark Ages, large-scale poultry farming disappeared in

CHICKENS

Europe and did not resume as a significant activity until the 19th century. It is said that the domestic chicken reached the Americas after Christopher Columbus discovered the New World in 1492, but it is possible that they arrived from Asia somewhat earlier.

In the 19th century, the popularity of the chicken was boosted in Europe by the interest in the bird shown by Britain's

Up until around 1960 in the United States, chickens were raised primarily on family farms, their chief purpose being to produce eggs, their meat being regarded very much as a secondary product.

DOMESTICATION OF THE CHICKEN

Queen Victoria. The Victorians, never slow to embrace new fashions and ideas, began to form breed clubs and societies, organizing exhibitions and shows in which prize specimens were sold for huge sums of money. It is said that Victoria banned the sport of cockfighting in 1849.

Today, many breeds are kept for their meat and/or for their eggs, with a smaller number that are prized for their ornamental qualities. As with many other forms of intensively bred domestic animals, size and appearance vary considerably, as do plumage colours and

BELOW: Traditional painted roosters on sale at Belém, Lisbon, a place from which many of the great Portuguese explorers embarked on their voyages of discovery.

OPPOSITE: Decorated Easter chicks from Mexico.

body shapes. Although some chickens closely resemble the jungle fowl, others look much more like the species of gamebirds known as francolins. Smaller chicken breeds may weigh only about 1.1lbs (0.5kg), even less in the bantam forms, whereas others tip the scales at over 14lbs (6.5kg). In 2003, it was estimated that the number of individual chickens in the world was more than 24 billion, signifying there are more chickens in the world than any other type of bird.

Chickens as symbols and in religion

Chickens have long been used in some cultures in religious rituals and ceremonies and as sacrifices to the gods. The ancient Greeks did not normally use the chicken in sacrifices, although they ate its eggs, and they even believed that lions were afraid of cockerels. The chicken was sacred to Athena, the goddess of warfare and wisdom, and was also a fertility symbol connected with Persephone, the queen of the Underworld; a symbol of Eros, with connotations of desire and love; and symbolized productivity in relation to the god Hermes. In many of the folk tales that originate in central Europe, the devil is thought to flee from the sound of a crowing cock.

In the traditional Jewish ritual known as *Kapparot*, performed at dawn or early in the morning before Yom Kippur (the Day of Atonement), a chicken is rotated and the following words intoned: 'This is my exchange, this is my substitute, this is my expiation. This chicken shall go to death and I shall proceed to a good, long life and peace.' The chicken is then given to the poor as food.

The chicken is a symbol of great religious significance in Indonesia, as part of the Hindu cremation ceremony, although in this instance the bird is not killed. Instead, it is kept tethered by a leg throughout the ceremony as a way of

LEFT: St. Peter, whom Jesus at the Last Supper predicted would deny all knowledge of Him before the rooster heralded the new day.

OPPOSITE LEFT: A painting of a cockerel from the island of Taiwan.

OPPOSITE RIGHT: Detail of a Chinese calendar that includes a rooster. The last Year of the Rooster was 2005.

ensuring that any evil spirits present may enter the chicken instead of the relatives of the deceased attending the funeral. After the ceremony, the chicken is returned home to continue its normal life.

In the Bible's New Testament, Jesus foretold, during the course of the Last Supper before His Passion, that before the cock crowed that day, Peter would deny knowledge of Him three times; because of this the cock came to be regarded both as a symbol of betrayal and of vigilance.

The rooster is one of the 12 animals that feature in the Chinese calendar. Reflecting, perhaps, its self-regarding nature, people born in the Year of the Rooster are thought to be somewhat self-absorbed and fond of the limelight. They are also well-groomed and tend to dress to impress, being keen on entertaining and with the ability to cope well with changing circumstances. People born in the Year of the Rooster are also described as confident and independent, and their most suitable occupations seem to be those that involve meeting or communicating with others, such as hairdressing, journalism, public relations,

teaching, newsreading, the police and the military. Roosters are believed to be most compatible with people born in the Years of the Ox, Dragon and Snake, but incompatible with those born in the Year of the Goat.

In Chinese culture the chicken is rarely used as a religious offering, but in Confucian weddings may be treated as a substitute for someone who is ill or otherwise unable to attend the marriage ceremony.

CHAPTER THREE
KEEPING CHICKENS

Here we are going to examine what it's like to keep a single chicken, perhaps two, as a pet either for a child or even as a companion for a grown-up. We will also look at the pros and cons of keeping a small flock of chickens as a way of having your own supply of fresh eggs or even meat. Here, some of what is said about keeping a single chicken applies equally to keeping a flock, while much of the advice on feeding and general care is also appropriate to both.

First of all, and this applies to one chicken or several, the keeping of chickens is not permitted everywhere, so it may be best to check first with your local authority. In any case there are liable to be some restrictions, and you will most likely find that your chicken or chickens won't be allowed to range free but must be kept in a suitable enclosure;

Chickens make good pets for children, provided they have been taught to handle them correctly.

there may also be some kind of ruling as to how close to human dwellings such a structure may be sited.

Pet chickens

Chickens have been kept as pets for centuries, and choosing the right one for the purpose often yields a surprisingly interesting and friendly companion. As with most other types of living things, some individual chickens are tamer than others, even within a breed which is generally known to be well-disposed toward human beings. And it is certainly true that some breeds make better pets than others. In the descriptions of the breeds of chicken that follow (*see Chapter Four*), there is usually some indication of the types generally thought to make good pets.

On the whole, breeds suitable for this purpose are those that are relatively light in weight (large chickens are hard to pick up, especially by children). It is also sensible to choose a fairly hardy type and one that has no fancy or elaborate feathering, especially around the feet, since they are liable to cause

Many enjoy seeing their chickens pecking about in their back yards, and a colourful rooster is a particularly fine sight.

problems in some cases if they are neglected. Bantams (usually smaller versions of 'standard' breeds, although sometimes breeds in their own right) can often be a good choice. Avoid male birds, since they can be aggressive by nature. Some chicken breeds live longer than

FAR LEFT: Chicken feeders allow just enough food through, keeping the rest dry and rodent-free.

ABOVE: There are many varieties of pedigree chickens, all with different characteristics. This is a Silkie hen.

81

CHICKENS

Free-range chickens like nothing better than to be given the run of a garden, where they can peck around in their search for insects that may be incidentally harmful to existing plants.

others, and this may also need to be taken into account.

Like many other kinds of animals, including other bird species, individual chickens each have their own traits and characteristics – let's call it their personality – and these features will become apparent as you study and get to know your pet. If handled kindly from an early age, chickens can become quite affectionate, and one hen we know had the habit of coming up and resting her head on her owner's leg, as if asking for her head to be stroked.

A traditional little brown hen. Chickens aren't the best of flyers, so special care must be taken to ensure their safety from predators such as foxes, and from local cats and dogs.

Another aspect of keeping chickens is that many highly ornamental varieties now exist, with features such as unusual coloration, feathering, and so on. In the event of choosing such a bird, showing them can be an interesting and rewarding pastime, and visiting exhibitions and shows can open up a whole new field of interest. But remember that some ornamental breeds may be less hardy and require more care to keep them in peak condition than other breeds.

As pets, chickens can be thought of as being reasonably 'low maintenance', which doesn't mean they should in any way be treated in a cavalier fashion in terms of their welfare. Some chickens are hardier than others, as already mentioned, but they all need warm, dry and safe places in which to roost at night, and somewhere where they can shelter from the heat of the sun or from inclement weather during the day. Given suitable conditions and the right food, chickens will thrive, but their health will soon begin to suffer if they are neglected.

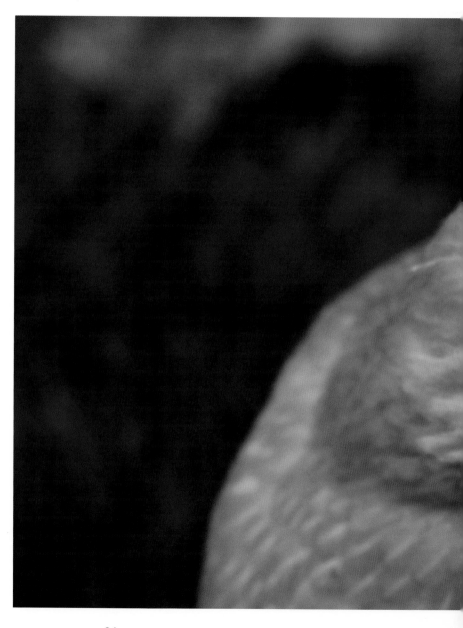

KEEPING CHICKENS

The fox is probably the greatest threat and will go to great lengths to get to your chickens, so great care should be taken to protect them at all times, and particularly at night.

One of the greatest hazards facing chickens in a domestic setting is that of predators, and other pet animals, such as cats and dogs, will certainly take their toll given the chance. Wild animals such as foxes, squirrels, raccoons and other prowling creatures are a real threat, and extremely upsetting though this may be, a fox will do more than kill a single chicken, and is liable to devastate an entire flock once it gets into an enclosure. Once considered more or less nocturnal, foxes of late have become bolder and more opportunist, being commonly seen by day in many places, so keeping chickens securely fenced at all times is good practice, particularly if they are allowed to range free.

Rats and other rodents are also a problem in that they are attracted to places where there is a generous supply of spilled food, and beneath a chicken roost would be the ideal place for a rat or mouse family to set up home. One of the biggest threats is Weil's disease, which is carried in the rat's urine and can enter the

CHICKENS

Good poultry houses protect chickens from the elements, injury and theft, and must also provide a stable environment in which the birds can feel comfortable during the day and at night, are protected against potential predators, and are provided with secure nesting boxes.

body through cuts and can also contaminate water. If left untreated the disease can be fatal, which is one reason why gloves should always be worn when cleaning out chickens. If you suspect rats are around – and you may even see them

as they can be quite bold on occasions – put down some poison or traps or, better still, get professionals in to deal with the problem. It goes without saying that any such measures must be implemented with due regard to the safety of children, pets, and the chickens themselves.

A secure place for chickens

Many types of proprietary chicken housing systems are available, and there are plenty of companies around which can supply the items or even design and erect 'bespoke' housing especially for you. Have a look on the Internet. In

Chickens must be kept in clean and secure conditions in a stress-free environment, where they are able to grow, sleep and lay eggs in comfort. They also need an outside space where they can take exercise, peck around at will, and take frequent dust-baths.

appearance, many typical proprietary systems are fairly similar to the type of structure often used for keeping pet rabbits or guinea pigs outside. They are constructed of a wooden frame, which may be square or rectangular, usually with a side door for access. Other types may be triangular in cross-section, but in all cases the base of the run or pen is open to allow the bird to scratch and forage on the ground. One end of the pen includes the roost, which is a solid structure with a secure door and a waterproof top.

Unless you intend to let your chickens range freely, then an alternative system that works well is to keep them in a large, strongly-made and permanently sited pen – usually made from wire mesh

LEFT: These chickens are living within a large, well-fenced pen where there is space for foraging and a place to which they can retreat at night.

ABOVE: An ideal night-time environment is a barn where there are plenty of high places where chickens can roost.

attached to a wooden frame – and to include within it or as part of it a sturdy, solid-sided but well-ventilated roosting place. The outer pen must have sufficiently high sides to prevent the chickens from escaping, and if possible be high enough to discourage cats and other animals from getting in too easily;

5ft (1.6m) is the minimum for an open-topped run. A safer method is to have a relatively high-sided pen with the top covered with mesh as well, making it completely secure. Such an enclosed system is regularly used by keepers of cagebirds that are housed outside. Sometimes the best solution to prevent

marauding foxes is to install electric fencing or electrified netting.

Once erected, although the fencing around your pen may seem in good order, check it regularly for signs of holes along the bottom; a fox may be trying to dig beneath the fencing to get in – a common and effective method used by them when

OPPOSITE: *Always ensure chickens have plenty of fresh water to drink.*

RIGHT: *Chickens are sometimes allowed to roam around gardens and vegetable patches at will, where they do little damage and may even be beneficial in controlling pests.*

trying to get into suburban gardens. It is best to ensure the bottom of the fencing extends a reasonable distance below ground (about 12in/30cm) to deter burrowing. Also, make sure the gauge of the mesh is correct – if the holes are too large your chicken may get its head caught while trying to peck at something outside. Holes that are too large will also allow the opportunist paws of a would-be predator to grab an unsuspecting chicken; 0.5-in (about 13-mm) mesh is usually recommended. As a general rule, if you are to put your chickens in a covered run, with occasional access to the garden or some other free-range spot, you should allow a pen area of about 4sq ft (0.4m^2) per bird, but if they are to be confined at all times, then allow more space per bird, in this case about 10sq ft (0.9m^2)

Any pen should obviously be positioned on suitable, well-drained ground. A roughly grassed area or one

CHICKENS

This chicken is living in a well-fenced, permanent outdoor pen, well-protected by high and stout wire fencing that has been sunk at least 12in (30cm) beneath the ground to deter burrowing predators; laying the buried part of the fence at a slightly outward-facing angle acts as an added deterrent. If there are doubts concerning the security of the pen, then the birds must be securely shut up in a chicken house before dark and released again early in the morning after the sun has risen.

KEEPING CHICKENS

LEFT: To be comfortable at night, a chicken needs a dark place and a high perch on which to sleep.

OPPOSITE: The door of this stout chicken house can be pulled down and securely closed at night.

that has soil and a few bushes is ideal: the chickens will soon turn it into a suitable habitat by scratching around in the earth, pecking at the foliage, and generally making themselves at home. Tree stumps or other suitable objects scattered around will give them something that they can fly or hop onto, and will add interest to the environment. Like many birds, chickens like to dust-bathe, which helps to remove lice and other parasites and scours the feathers clean. Therefore, in at least a part of the run there should be an area where the chickens can dig themselves a dust hole. If no such area is available, then a suitable, low container should be provided, filled with a mixture of ashes, dry soil and sand.

Like the pen, the chicken house or roost can be purchased ready-made, or you can erect one yourself, and there are plenty of books and websites that provide instructions. A typical roost is set slightly off the ground with a short ramp leading

OPPOSITE: This simple but secure arrangement is fine for a warm climate.

ABOVE: A portable chicken house sited within a well-fenced and maintained pen.

to the access door. It should be dry, draught-free but well-ventilated. The floor of the roost should be covered with clean straw laid on a thin layer of dust-free wood shavings, about an inch or two deep, which are readily available from pet supply stores and which should be changed regularly. Keep an eye on the 'bedding' to check for droppings, which must be removed every day. A removable droppings tray placed beneath the roosting poles will help to ensure that everything is kept clean and healthy. The soiled bedding makes good compost, so nothing need be wasted. Make sure the roost is sited where there is shade, otherwise it may get uncomfortably hot inside during the summer months.

Allow an area of 1.5cu ft (0.04m^3) per bird in the enclosed space of a roost, and make sure that it is sturdily constructed with a strong base. Once the chickens are safely inside for the night, the access door to the coop should be shut securely with a bolt or a similar arrangement; if a predator such as a fox should get into the outer pen, the roost must be sufficiently robust to prevent entry here as well. The small wooden pivoting bar, so often seen on the doors of animal houses, is not sufficient protection against a determined animal like a fox, which will inevitably knock the bar down when scratching to get in and thus allow the door to open.

Chickens, being birds, have a natural urge to roost off the ground at night for security like others of their kind. Therefore roosting perches should be provided so that your chicken or chickens can rest snugly off the floor. Perches should be about 2in (5cm) wide for standard breeds but a little narrower for smaller birds. Allow 10in (25cm) across for the bird, and the same amount of space between roosting poles if you have

more than one chicken. If you do install several perches, it is best to have them graded slightly in terms of level, so that the birds do not all roost at exactly the same height. An inch or two is sufficient; don't set them so high that the birds have difficulty getting up to them. Roosting perches should be constructed of timber that has had its bark removed, as cracks and spaces left by fragmenting bark may harbour undesirable insects.

These chickens have the best of both worlds, having a substantial shed where they are secure and can roost at night, and a yard where they can roam freely during the day.

The roost should be cleared out weekly and fresh bedding laid down. Then, at least twice a year, but more often if conditions seem to dictate, the roost should be thoroughly stripped out

and deep cleaned. Everything should be removed, including bedding, nest boxes and any feeding containers. A safe cleaning agent should then be used to disinfect the walls, floor, roosting poles, and so on. Rinse everything with fresh water and allow to dry thoroughly before letting the chickens return. Fresh, clean bedding should be laid on the floor of the roost and in the nest box.

As long as the roost is well-ventilated but free of draughts (this is very important), there is normally no need to heat it in winter. Chickens will adapt to a falling temperature as the seasons change, and it is best that they

BELOW: Pet chickens often become tame enough to take food from the hand.

RIGHT: Chickens enjoy foraging for themselves in a companionable way.

PAGES 108 & 109: Suitable feeding methods.

should not be faced with the fluctuating temperatures that arise by going from a warm roost to a cold outside run. If the power should fail in a heated roost, this can also cause problems – even death – for birds that have been accustomed to warm conditions at night.

Food and water

Fresh drinking water must be available at all times when the chickens are in the pen. Under normal circumstances it is not necessary to provide water for them in the coop or roost as long as the birds are let out into the pen each morning to

access drinking water for themselves. Various types of water dispensers are available, such as fountains which automatically fill up a trough with fresh water. Whatever type of water supply is used, ensure it is fixed at the recommended height and in accordance

with the supplier's instructions. Don't use a water dispenser that might be fouled by the chickens hopping onto it. In cold weather, make sure that the water supply, or the reservoir if one is fitted, is not allowed to freeze; chickens won't be able to survive for long without water. To ensure it remains ice-free, water can be brought into a warm place overnight and returned to the run in the morning.

Being omnivores, chickens enjoy eating suitable kitchen scraps as a supplement to their regular pellet-and-grain feed and the food items they are able to obtain for themselves by foraging.

KEEPING CHICKENS

them getting wet. A little grain can also be added, such as wheat or corn, to augment the diet if preferred, which can be scattered around the pen area; most

The wild relatives of the domestic chicken enjoy a varied diet, ranging from seeds and other vegetation to insects and even small lizards. Domestic chickens are also omnivorous in their eating habits, and have even been known to catch and devour the odd mouse that gets into the enclosure! However, for those beginning to keep chickens, it is advisable to base the feeding regime around one of the good-quality proprietary complete pellet feeds that are available; this will ensure that they get a balanced diet of carbohydrate, fat and protein, vitamins and minerals (such as calcium needed for eggshells). Use a commercially supplied dispenser or hopper for the pellets to stop

Broody hens need stress-free environments and clean and comfortable conditions if they are to incubate their eggs so that hatching of the chicks can take place.

authorities advise this should be offered in the afternoon, with the pellet food being given in the morning.

Grit is essential for many birds, not only chickens. Because they are without teeth, the abrasive properties of grit are needed in their crops (part of the digestive system) to help them grind up their food. Some grit should be present in good-quality proprietary feed, but it is worth providing a small container with an extra supply so that the chickens can take as much as they need.

KEEPING CHICKENS

It is fortunate for those of us who enjoy her eggs that hens, given a little care and encouragement, are able to continue laying even when no cockerel is present in the flock. A male will be needed, however, if the eggs are to be made fertile and eventually hatch to produce chicks.

Chickens also appreciate a little fresh greenery, so let them get on to grass as often as is possible, or offer them plants to eat such as chickweed. Chickens are also happy to eat leftovers from the kitchen, but not of course as a substitute for properly designed chicken feed. They will eat fruit, cabbage, vegetable peelings, and so on. Do not, however, offer them anything strong-tasting such as onion, garlic or spices; neither should they be given citrus fruits (such as oranges and lemons) nor any food in the process of 'going off'; chickens shouldn't eat rotten food any more than we should ourselves; moreover, the likelihood is that it will be rejected and will be left to rot still further, eventually attracting vermin and flies.

Laying eggs

Even if you are keeping a single hen as a pet, she will reward you with eggs if you encourage her to lay. A nest box should be provided, raised a few inches off the ground within the roost; it should be lower than the highest roosting pole, however. Position the nest box in a quiet, dark place to help her feel more secure. Again, there is plenty of advice available in books and on the Internet about how to create the ideal conditions for successful egg-laying. For example, it is possible to buy dummy eggs made from stone that encourage first layers to sit on their eggs without crushing them. When I was young I used to have great fun dropping these eggs in front of people who didn't realize they were not real, just to see the expression on their faces! The bottom of the nest should be lined with a little clean straw or pinewood shavings to help prevent the eggs from breaking when they are laid. (Despite all these measures,

LEFT: Little bantams are highly decorative, and many are kept simply as garden pets.

OPPOSITE: Some breeds of chicken are more amenable than others and make better pets where children are concerned.

CHICKENS

you will inevitably find broken eggs in the roost from time to time.) Contrary to popular belief, a chicken will still lay eggs whether a cockerel is present or not. The only difference is that without a cockerel around to fertilize them they will never produce chicks.

Choose your chicken according to the purpose for which it is required, either for the fact that it is an excellent egg-layer, it produces tasty meat, or it has the friendly disposition that will make it an ideal pet.

KEEPING CHICKENS

Some chickens are happier roaming over a large area, while others quickly become accustomed to a smaller space.

Obtaining a chicken

Once all the necessary accommodation and equipment is in place you can begin to think about obtaining a bird and introducing it to its new home. At this point it is more than likely that you have reached a decision concerning the type of chicken you want and why you want it; in other words, will it be essentially for eggs, will it eventually be for eating, will it be for showing, or will it simply be kept as a pet? To help you make up your mind, the breed descriptions in Chapter Four will provide much information about different types of chickens, such as size and colour and their temperament and hardiness. However, the more you can find out beforehand about your potential purchase from other sources, including breeders and the Internet, the better. Seeing the actual breed in the flesh, moreover, is the best way of ensuring you will be ultimately satisfied with your choice.

If you wish to keep chickens primarily for the quality of their eggs, then a commercial hybrid is the best choice. If you would prefer the eggs to be

KEEPING CHICKENS

a particular colour, then go for a pure breed. Wyandottes and Rhode Island Reds, for example, are among the breeds that lay mid-brown eggs, Leghorns are among those that lay white eggs, while Sussex hens lay tinted eggs. Welsummers and Barnevelders lay dark-brown eggs, and Araucanas lay blue eggs. One of the drawbacks with hybrids is that they do not have the distinctive look of some of the more spectacular breeds, but in their favour is the fact that they are cheaper to buy, come vaccinated against disease, and are excellent layers, some producing well in excess of 300 eggs in a year. Hybrids, of course, are the mainstay of the commercial industry for both meat and eggs.

Most breeds have their own clubs, whose members can put you in touch with reputable local breeders in your area who can sell you a pullet (a young hen under a year old). It is always best to visit the premises of the breeder from which you intend to buy, so that you can assess

It can be a rewarding experience to give a new, less restrictive and natural home to an ex-battery hen, although it may take a little time for her to settle down.

the conditions under which the birds have been raised. Talk to the breeder and explain your reasons for wanting the breed you have chosen and try to obtain as much information as you can about it – how to care for it, any special traits it may have, and so on. Becoming a member of the breed club for the chicken you have chosen is a good way of obtaining more information. Another way, before you make a purchase, is to visit a poultry show or an event such as a country fair, where various breeds can be viewed.

Another route is to buy fertilized eggs and place them in an incubator until they hatch. This is particularly exciting for children, and there is the added benefit of seeing the chick at the very earliest stages in its life. If you are planning to try this, you must ensure that all the necessary equipment is in place before you obtain the eggs. There are plenty of books and websites that tell you how to set up an incubator, what the correct temperature should be, and what to do once the chicks have hatched (for

example, they need to be fed on chick starter feed for about six weeks, followed by a pullet feed). It is likely that most sources of information will also warn you to make sure that the chicks are kept safe from another form of predator – in this case the pet cat or dog, which may be tempted to eat a small chick sitting in a warm box in the house!

Ex-battery hens, which have reached the end of their commercial life – usually at about 72 weeks – are fast becoming another choice for the amateur chicken-keeper or for those who want a bird or a number of birds that they can keep on a small scale. There are several organizations that specialize in rescuing these hens and offering them to people willing to give them a new life. It can be an initially sad yet ultimately rewarding experience to take on one of these birds. Some of them will be traumatized and even in poor condition to begin with, having poor feathering and may even need to regrow their beaks. But just as with other rescue animals, such as cats and dogs, with the proper care and

Taking a bird that is very tame, teach your children the correct way to pick up and hold a chicken.

attention there is no reason why they should not in time return to their full vigour and fine appearance.

How to handle a chicken

Like most animals, there is a right and a wrong way to pick up a chicken, for one that is dropped or mishandled, though unintentionally, is likely to become wary and apprehensive even of being stroked, and won't make a good pet as a result. For this reason, it is advisable that young children are taught the correct way to handle a chicken right from the start. Remember that large breeds don't always make good pets; the urge to pick up and cuddle is enormously strong in children, but large animals don't fit into small arms!

Before you pick up a chicken, it is often useful to let it get to know you first. Try offering a few tasty morsels, or approach very slowly and attempt to touch or stroke the bird.

To pick up a chicken, place your dominant hand (the one you write with) on the middle of its back. Use the thumb and forefinger of that hand to gently but firmly secure the wings so that they can't flap. Now, using your other hand, enclose one of the chicken's legs with your thumb and forefinger, and the other with the forefinger and middle finger of the same hand. Now lift the chicken up, using the heel of your hand and your wrist to support the underside, keeping your dominant hand on the chicken's back in the meantime. Now you can hold the chicken close to your body to prevent it from wriggling; it's best to sit down with the chicken at this point. After a time, the chicken will come to realize that this activity on your part results in its movements being somewhat curtailed, and as time goes on it should cease to struggle. Soon you should be able to pick it up without any fuss, especially if you talk to the bird gently and stroke it while it is in your arms.

Keeping chickens healthy

Chickens are pretty robust and healthy most of the time, but they can succumb to illness and disease like any other animal. Chickens kept in flocks can quickly transmit diseases to others, which will compound the initial problem. If detected early enough, however, most ailments can be treated. Before this happens, however, it is worth acquainting yourself with details of a vet in your area who is skilled in avian (bird) medicine. Ask your local

This healthy rooster is full of vitality.

vet if he can provide such a service, and if not, ask him to recommend another who can. Your local authority or pet rescue centre may also be able to offer advice, since they may themselves call upon veterinary services from time to time. Symptoms of disease can manifest themselves in various ways, so it is sensible to check each chicken daily to ensure they are not presenting any early signs. Look in particular for the following:

- Listlessness or loss of appetite – does the chicken seem to be behaving differently from normal?
- Poor feather condition – feathers should appear sleek and 'well-groomed' in a healthy bird. Do not confuse poor feather condition with moulting, which all birds undergo to replace their feathers as part of the natural cycle. Egg-laying will also be curtailed during the moulting period.
- Any obvious change in the pecking order; if you have more than one chicken, is the one in question being bullied or dominated by one normally lower in rank?

Check your chickens regularly for signs and symptoms of disease.

- Sneezing.
- Parasites on the feathers or skin, or a generally 'mangy' appearance.
- Unusual-looking stools (droppings) and/or the appearance in the stools of worms; normal stools are brownish with a white 'cap'.

It must be stressed, however, that these are not the only signs of illness, and if you have any other reasons to suspect that not all is well, expert advice should be sought as soon as possible.

Many diseases may be brought about by poor hygiene regimes in the coop, so make sure that the cleaning procedures described earlier are followed at all times.

In addition to dealing with illnesses, there are a number of measures that should be taken to ensure your chickens remain healthy. These include preventing worms or dealing with them once they are detected. There are several types of parasitic worm, of which the most common is the roundworm. Chickens afflicted with these can suffer weight loss, feather-fluffing, diarrhoea, and in severe cases even death. Boggy ground, that attracts earthworms (one of the carriers of roundworms), can be a cause of the problem. Tapeworms are less commonly problematical but are still a cause for concern, and similar symptoms

as those shown by roundworm infestation are often seen. In cases of worm infestation, or to prevent worms, an anti-worming treatment is advised. This should be carried out at least twice yearly, or according to your vet's instructions. During the treatment, and for a period following it, none of the affected chicken's eggs should be consumed.

The beaks of birds grow continually, as do our own fingernails, this constant growth being an important factor that repairs normal wear and tear. Chickens that range freely tend to keep their own beaks in trim through the abrasive action of pecking around in the soil. However, it is possible that the upper mandible (the top part of the beak) will sometimes overgrow the lower one if the chicken is not allowed to feed in this manner – for example, if the food is too soft and the conditions for general pecking and foraging are not met. If the beak does overgrow in this way, it must be trimmed back, otherwise the chicken will not be able to eat and drink normally. Experienced chicken-keepers often do this job themselves, but initially it is

Beaks and claws grow quickly and should be regularly monitored and trimmed if necessary.

advisable to have a vet or similarly qualified person do it for you so that you may see for yourself how it is done.

The claws of a chicken may also need trimming, since they grow continually in the same way as the beak.

Again, in ideal conditions, all the scratching at the ground which a chicken does in the normal course of finding food

Check chickens regularly for lice and mites, consulting your vet if evidence of their presence is discovered.

will tend to keep the claws at the correct length, but if they get too long – usually as a result of walking on soft litter – then they will need to be cut back, otherwise foot problems can ensue. Like the beak, this is a job you can do yourself, but to begin with, a 'lesson' from an experienced person is required.

Some chickens are born with beak defects, such as one that is twisted to the left or right instead of growing straight. There may also be a visible gap between the upper and lower mandibles when the mouth is closed. Such conditions can affect the bird's ability to eat properly, and special foods may be required. Unless caused by an injury, defects of this type are normally genetic in origin, and if you intend to breed your chickens it is best not to use such birds in order to avoid passing the condition onto the next generation.

External parasites, such as lice and mites, can also be a problem. Lice are easier to spot than mites, since they are bigger; they may be seen around the bases of the feathers and their eggs may be seen visibly attached to them. Look

CHICKENS

particularly around the vent and under the wings where the softer downy feathers grow. Dust-bathing helps chickens get rid of lice naturally, but special treatments are also available to rid them of these troublesome parasites.

There are several types of mite: red mites do not actually live on the chicken, but lurk in crevices in the roost. At night, when the chickens are resting on their perches, the mites come out to feed on the hosts' blood. The irritation they cause can make the chicken pluck out its own feathers or desert its nest if brooding, so a complete eradication programme is needed to deal with the problem. As well as treating the birds, the infestation must be removed from every likely place in the roost by washing (preferably with water under high pressure) and then treating all surfaces, including perches, with a poultry-safe disinfectant. Another type of mite infesting chickens is known as the northern fowl mite, and the bird should be treated to eradicate these. Another mite-related condition is scaly leg, which is often brought about by keeping birds in damp living conditions. The mites burrow into the scales on the chickens' legs, which then become inflamed and painful. If left untreated, the mites' excreta builds up and causes the scales to lift away from

To avoid disease, keep coops clean and disinfected and regularly change soiled bedding material.

the legs, in which case your vet should be asked to deal with the problem.

Various other diseases may be encountered, such as coccidiosis, which is a minute single-celled parasite. It is usually present in low levels in chickens (chicks acquire a measure of immunity to the parasite through exposure as they develop), but poor housing conditions, such as overcrowding and lack of ventilation, can exacerbate the condition. Young chickens from about three to six weeks old are most susceptible to the disease, but older birds can also suffer. Giving the chicks a feed containing an anti-coccidial agent will help to reduce the risk.

Mycoplasma is a respiratory disease affecting birds, usually those that have already been weakened by a virus, since the condition is often already present in the bird in low concentrations. Marek's disease is a virus that often affects domestic chickens, although it is rarer in commercial stock since they are inoculated against the disease when they are a day old. Particularly susceptible are Silkies and Sebrights.

CHICKENS

A compacted crop is a common problem with chickens. As we saw earlier, the crop is part of the bird's digestive system, in which food taken by the mouth is stored before passing to the gizzard, where it is ground up prior to being digested. Somehow, the food in the crop does not always pass down to the gizzard and instead becomes blocked, indigestion, obstruction or ill health being the main reasons. Offer the chicken some warm water to drink while at the same time rubbing the crop gently with your hand to loosen the compacted material. Then, holding the chicken with its head held slightly downward, gently squeeze the crop, which should force the liquid to run out. Now offer more water and then prevent the chicken from feeding for 24 hours. Again, this technique is one that is best done after having watched someone perform it first.

One of the conditions that often afflicts chickens kept as pets is obesity, brought about by the same reasons that cause the problem in humans, i.e., too much food, and of the wrong type, and not enough exercise. Chickens are a little like some breeds of dog in their eating habits, tending to wolf down whatever they are given, especially if it is kitchen scraps. Poor egg-laying and infertility are just two of the problems connected with an overweight chicken. To check for obesity, feel for the breastbone; it should be possible to detect it beneath a reasonable layer of fat. If the breastbone feels sharp, it is likely that the chicken is in fact underweight, but if it is difficult to feel it at all, then the likelihood is that the bird is carrying too much weight. To avoid overfeeding, make sure no food remains by the time the bird is ready to roost, but if there is food left over, try cutting it back a little. As a rough guide, each adult bird should receive about 4oz (113g) of food per day, given that they may well supplement this amount with extra food obtained from foraging.

Bird or avian flu

A devastating form of this disease has been much in the news over the last few years. Like human beings and other

species, birds are also subject to this disease, the most contagious of which are the H5 and H7 strains. The virus is usually transmitted to domestic stock from wild migratory birds, such as geese, which are natural carriers of viruses. Domestic poultry can suffer from several forms of bird flu, which may be relatively mild or much more virulent. Symptoms of mild bird flu include ruffled feathers, lowered egg production, and mild respiratory problems.

The form of the virus that is causing most concern, however, and which has resulted in the need to cull enormous numbers of poultry in the Far East, is the strain known as H5N1. Unfortunately, humans can catch the disease from infected poultry if they are in close contact with the birds. The virus is excreted in the stools, which then become dry, break up, and are then inhaled. In birds, the symptoms of this deadlier form are a rapid decline and death – up to 100 per cent of all infected stock within 48 hours according to World Health

The pecking order will need to be re-established when adding further chickens to an existing flock. This is facilitated by providing added distractions and eliminating overcrowding.

Organization figures – which includes haemorrhaging from organs. The disease can be equally dangerous for human beings and has already killed several people who became infected via contact with poultry. However, it is far more worrying that the virus might spread through contact with infected people travelling across the world, leading to an influenza pandemic.

Clearly, if you have any suspicions that your chickens have any form of bird flu, the vet must be contacted immediately, so that the appropriate biosecurity measures can be put into place as quickly as possible.

If a chicken has died through contracting a suspected contagious disease, such as bird flu, its body will be disposed of by the relevant authorities, who will need to do a post mortem to establish the exact cause of death. Depending on the outcome, they may instigate other measures of control to

Introducing new birds to an existing flock can be something of a problem, giving rise to frequent shows of threatening behaviour; but the pecking order should become re-established within a couple of weeks if all goes well.

avoid spreading the disease. Chickens, however, can also die of natural causes, with old age or an attack by a predator, such as a fox, being the most common reasons. Children are especially affected by the death of a pet, and giving the chicken a 'proper' burial will help to console them in their grief. The best course of action is to bury the chicken in the ground, placing it in a hole several feet deep to prevent it from being dug up by animals.

New arrivals and the pecking order
Once you have a few chickens you may decide to add more to your flock, and this is a process that needs to be handled in the right way. We have already seen that chickens live according to a hierarchical system, known as a pecking order, and healthy chickens will live quite happily under this self-imposed regime once it has been established. Once newcomers are introduced into this nice and tidy arrangement, however, the whole system is temporarily thrown into confusion, and the birds will need to re-establish their relative places within the flock. Things will usually settle down again within a week or so, but in the meantime much quarrelling, bullying, and general unpleasantness is likely to ensue. There

are things you can do, however, to alleviate the turmoil to some extent.

The bullying of newcomers will be more severe if the existing chickens have little to do, but if they can be distracted from this activity so much the better. Introducing some additional features into the run, such as a few large tree branches with twigs, will give the newcomers somewhere to hide or escape the attention of the aggressors, while providing extra food sources, such a grass cuttings, dead leaves, some weeds or some kitchen scraps, will also give the hens something more interesting on which to focus their attentions. Hanging a cabbage just out of their reach will also give them something else on which to concentrate. If possible, letting the chickens out to range freely is another good way of reducing the tension, allowing them to become accustomed to the newcomers while they concentrate on foraging. Do not allow the newcomers out to range freely for a few days until they have recognized that the run and the coop are their new home.

Should your chickens develop bare patches of skin throughout the year, it is likely that they are pecking one another – usually a sign that they are under stress. Since contented chickens do not usually indulge in this behaviour, it is important to investigate the source of the stress so that it can be removed. But beware that in the worst cases, a chicken may even resort to eating another. Check to see that there is sufficient space (*see guidelines mentioned earlier*) in the run and in the roost. Are the chickens competing for too few feeders or water containers? Are there enough nest boxes and roosts to go around, and are they positioned sufficiently distant from one another? Are the chickens infested with parasites? If none of these factors appears to be the cause of the problem, then it is advisable to consult your vet, so that he can make his diagnosis and suggest a possible solution to the problem.

To avoid aggressive behaviour and relieve stress in the flock, make sure your chickens are not overcrowded and have plenty of space where they can wander at will. Check also that sufficient feeders and water containers have been provided, so that one more excuse for picking a fight is eliminated.

CHAPTER FOUR
BREEDS OF CHICKEN

Today there are over 70 breeds of chicken, and many of them are available in a range of varieties, including smaller bantam forms. The physical traits used to distinguish breeds are size, plumage color, comb type, skin colour, number of toes, amount of feathering, ear-lobe colour, egg-colour, and place of origin. There follows a description of many of these breeds and their varieties.

APPENZELLER

The Appenzeller is a Swiss breed, known for hundreds of years and named after the region in Switzerland where it was

RIGHT, OPPOSITE, PAGES 156 & 157: Silver-spangled Appenzeller Spitzhaubens. A common variety elsewhere, these are rare in North America and are officially recognized neither by the American Poultry Association, nor by other US breed registries.

developed. The breed comes in two versions: the Barthuhner and the Spitzhauben, the Barthuhner being the more robust-looking of the two. The Barthuhner is also bearded and has white ear lobes; it also has a rose comb rather than a crest on the head.

Appenzellers have tight plumage, with striking markings and wide-spread, upright tails. The eyes are dark brown. Silver-spangled Spitzhauben varieties have silver-white feathers with black, while in the gold-spangled varieties the ground colour is golden-red. On the head

the unique crest is forward-facing – said to resemble the traditional bonnets worn by women in the Appenzellerland region – and there is also a split, horn-type comb. The wattles are long and fine and the beak is relatively large and powerful. The legs are devoid of feathering and blue-grey in colour. As well as the silver-spangled and gold-spangled varieties, there is also a black variety.

The Appenzeller is a very flighty and active bird, which seems to have a rather nervous disposition and is not recommended as a pet for beginners. The breed does not take well to confinement and needs plenty of space, and care should be taken of them in cold weather. They are able to forage well given the opportunity. The Appenzeller is a good layer and is capable of producing about 150 white eggs per year.

BREEDS OF CHICKEN

ARAUCANA

The very distinctive Araucana chicken originated in South America, named after the Arauca Indians of Chile and first mentioned in documentation dating from about the mid-16th century. It was introduced to Europe in the early 1900s, and the breed standard was set at a meeting of the Rare Breeds Society in 1969. Originally, the bird had a large, floppy pea comb, but through breeding the comb is now a small, irregularly shaped pea comb. There are no wattles. The feathers on the face are very thick, and on top of the head there is a small crest. There is an unusual wart-like feature on each side of the head where in other breeds ear lobes are usually seen. The birds are deep in the body with an upright stance and with a fairly long back, reminiscent of the Asiatic game type. They may be born with or without a tail; those without a tail are known as rumpless Araucanas, and this feature can occur in both sexes. Males weigh 6lbs (2.7kg) and females 5lbs (2.3kg). There is also a bantam variety.

Another distinctive feature is the bird's blue-green egg, the colour of which extends all the way through the shell. Hens tend to lay during the spring and summer months only, generally becoming broody and making good mothers. Araucana chicks are strong, growing fast and maturing quickly. They are considered to be dual-purpose birds. Araucanas are placid birds that do not mind being kept in pens, which should be moved around from time to time to give the birds access to fresh grass in addition to their normal feed.

The Araucana is available in a number of different varieties, including blue, black, cuckoo, lavender, gold duckwing, black-red, white and spangled.

AUSTRALORP

The Australorp is an Australian breed developed from Black Orpington stock that was imported into Australia from England in the 1890s and early 1900s. Here, the imported stock was crossed by local farmers with breeds such as White Leghorns, Minorcas and Langshans to improve the general quality of the Orpingtons. The original birds were called Black Utility Orpingtons, underlining the utility value of the new breed. There was some initial controversy concerning the name that should finally be given to this new breed, as indeed there was about agreeing a national standard for the bird between the various Australian states. The name 'Australorp'

OPPOSITE: The Araucana is sometimes mistaken for another breed, the Ameraucana, that was developed in the United States in the 1970s.

THIS PAGE: The Australorp: in black varieties, the feathers have an iridescent green sheen.

was penned by a poultry fancy institution before the First World War, another suggested name having been 'Austral'. But it was then suggested that the name should reflect the importance of the Orpington breed in the original development. By the early 1920s the

CHICKENS

name Australorp was in common use, at which time the breed was launched internationally. In 1929 the Australorp achieved Standard of Perfection status.

Hardy and docile, the Australorp is smaller and neater than the Orpingtons from which it derived, although it is still a substantial bird, and its soft black feathers have an almost metallic green sheen. The eyes and legs are dark, and the combs, cheek patches and wattles are red; the comb has five distinct upright points. The Australorp isn't a good flier, even though it is quite active and likes to range free, so fencing does not need to be too high to contain the bird. It is prized for both its egg-laying (light-brown eggs are laid) and its meat qualities. Males weigh around 8.5lbs (3.9kg), with females weighing in at 6.5lbs (3kg).

Although the Australorp is one of the best of the dual-purpose chickens, the breed's egg-laying abilities first brought it to the attention of the international community. In 1922–23, six hens set a world record by laying 1,857 eggs at an average of just over 309 eggs per hen over a 365-day consecutive period – and this at a time before the regime of intensive laying techniques and artificial lighting, so often in use today, had become established. The outcome of the birds' extraordinary performance meant that orders for the breed came in from all over the world, including the USA, Britain, South Africa, Mexico and Canada. Today, given the right conditions, a hen will on average lay about 250 eggs per year. The Australorp is a friendly bird and makes a good pet for children, although the smaller bantam variety is easier to pick up and hold.

BARNEVELDER

This is a fairly common breed of chicken whose name derives from the Dutch town of Barneveld. The Barnevelder is a hardy bird, bred to thrive in damp, windy conditions such as those often prevailing in The Netherlands, where it is among the most popular breeds. It is also a favourite in countries such as Great Britain, where a similar climate is often encountered.

The Barnevelder was created by selective breeding during the second half

OPPOSITE & LEFT: Barnevelder hen and chicks. The Barnevelder gained worldwide recognition and was exported to many countries because of its ability to lay approximately 180–200 large brown eggs per year.

of the 1800s, when Asian breeds, such as Cochins, Brahmas, Malays and Croad Langshans were crossed with local Dutch chickens. One of the resulting crosses was bred with Brahmas, and the resulting offspring were crossed with Langshans. Then, toward the end of the 18th century, some American chickens were introduced into the breed. In 1906 Buff Orpingtons from Britain had also been introduced, and by 1910 the breed was producing uniform colour and type, becoming established as the Barnevelder. The breed was imported into Britain in the 1920s but is also found in many other parts of the world.

A medium-sized breed, male Barnevelders grow to about 8.5lbs (3.9kg) in weight and females can achieve about 6.5lbs (3kg). The male has a prominent red comb extending from the base of the beak to the back of the head, and a longish red wattle. The skin around the orange eye is also red, and the legs are yellow. Hens have correspondingly smaller combs and wattles.

Several varieties exist:

Double-laced fowl This variety has a single comb and a distinctive, double-marked feather pattern. Typically the shawl and the back and tail of the cock are black, with greenish and brown feathering on the breast and wings. Hens

are browner overall.

Double-laced bantam This variety resulted from crossings between double-laced hens and Rhode Island bantam cocks, followed by other crossings with Wyandottes and Cornish.

Black bantam This variety was bred from black Barnevelders and black bantam Wyandottes.

White fowl Occasionally white individuals appear. These were mated with white Plymouth Rocks and white Leghorns to produce white Barnevelders.

Today the bird is kept as a dual-purpose breed and also for showing. A ready layer, it produces good-sized, brown eggs, and a reliable hen may produce as many as 200 eggs in a year. Not only is the breed a hardy one but it is also a good forager, thus requiring less intensive feeding than some other varieties.

While Barnevelders became famous for their dark-brown eggs in the first half of the 20th century, most birds appear to be in the hands of show-breeders today, resulting in less attention being paid to maintaining their productivity and the colour of their eggs. Instead, the focus is now concentrated on the birds' external characteristics.

Brahma

The name Brahma is derived from the Brahmaputra river in India, yet the breed was actually created in the United States from birds with well-feathered legs, known as Shanghais, imported from China. Development of the breed took place in America between 1850 and 1890. Brahmas grow and mature slowly, taking about two years to reach full maturity, so do not find great favour with commercial farmers. However, the commanding appearance of the Brahma, together with the striking plumage, make them popular with showers and fanciers. Brahmas are also elegant, stately and dignified birds, and their gentle nature and large size, together with the intricate plumage patterning, make them particularly attractive ornamental fowl.

Brahmas have broad, deep bodies, full breasts and long, powerful orange or yellow legs and feet covered with an abundance of soft feathers, giving them the appearance of having big, soft feet. The profuse feathering also stands the

The origins of the Brahma lie in India, where they were known as 'Grey Chittagongs', and believed to be closely related to the jungle fowl (Gallus gigantus) and the Cochin.

ABOVE & OPPOSITE: The American Standard of Perfection recognizes three Brahma varieties, light, dark and buff, the chickens pictured here being of the light variety.

wattles, features which again help the breed withstand cold weather. Males can achieve 12lbs (5.4kg) and females 9.5lbs (4.3kg) in the light variety, with males reaching 11lbs (5kg) and females 8.5lbs (3.9kg) in the dark and buff varieties.

Brahmas do not fly, and can be confined behind fencing about 2- or 3-ft (0.6–0.9m) high, although their large size means that quite a lot of space in which to roam around is required. The profuse feathering on the legs and feet means that dry conditions suit the bird best; the feathers can soon become clogged in wet conditions and mud –even faecal – balls may develop on their toes, which must be removed to prevent the bird from losing the claws or even the tips of the toes.

Because they are large and docile, Brahmas are reliable brooders, capable of covering a large number of eggs at any one time. The brown eggs are small, but chicks hatch strongly and grow quickly. Hens begin to lay when they are around six or seven months old, and will continue to lay throughout the winter, unlike some pure breeds. Brahmas will tolerate other breeds without causing problems, and may even be submissive toward them despite being the larger birds; even cocks tolerate one another. Brahmas are not noisy, and the cockerels are said to crow quietly!

breed in good stead in cold weather. The head seems disproportionately small for such a big bird, the face being smooth and free from feathers. The eyes are large and prominent, and the beak is short and strong. There is either a small red triple comb or a pea comb, and small red

BREEDS OF CHICKEN

CAMPINE

The Campine originated in the Campine region of Belgium several centuries ago, and was much prized for its white-shelled eggs. It is thought to be derived from another, slightly larger, Belgian breed, the Braekel, which is itself one of Europe's oldest breeds, originating in the 1400s. The Campine was introduced to Britain in the 19th century, but has never been especially popular in the US. This is a fairly small, attractive breed with close, barred feathering. Males are easily distinguished by their longer tail feathers, more showy neck feathers, and their larger combs. The combs can be prone to frostbite, and the birds themselves are not especially hardy, so adequate protection at night or in cold, frosty weather is required.

Campines have lively temperaments and prefer to be free-ranging, although they will adapt to being kept in more confined conditions. They are also inquisitive and alert and will take to the air readily. Not all individuals become tame. They are kept mainly for their ornamental value, favoured for their plumage and upright stance, although they produce a good number of medium-sized eggs given the right conditions. They are a non-broody breed.

The Campine has been bred in two colour varieties – golden and silver. In both varieties the ear lobes are white and the combs and wattles are red. The legs are slate blue.

CHANTECLER

The Chantecler is a Canadian breed originating in the province of Québec. It

LEFT: A silver Campine.

BELOW LEFT: A golden Campine hen.

was created to produce a chicken with good dual-purpose qualities, but also one which would be able to withstand the harsh Canadian climate. Thus, for example, the Chantecler has a small comb and wattles to help reduce the effects of frostbite on these exposed parts of the body. The breed came about as the result of two crosses: in about 1908 a male Cornish was mated with a female Leghorn, and a male Rhode Island Red was crossed with a female Wyandotte. Later, females from the first crossing were then mated with males from the second. The Oka Agricultural Institute in Québec first exhibited the breed, and it was given breed status in 1921. A second type, the partridge variety, achieved breed status in 1935.

The Chantecler has a slightly elongated body with a deep, well-rounded chest and with the tail held at about 30 degrees above the horizontal. The head is broad and short, the face is red, with reddish-brown eyes, and the comb and wattles are both small. The ear lobes are oval and off-white in colour. The short beak is yellow in the white variety, but

CHICKENS

The only native Canadian breed, the Chantecler was developed in the early 20th century at the Abbey of Notre-Dame du Lac in Oka, Québec.

horn-coloured in the partridge variety. Plumage is profuse on the neck where it rests on the shoulders. The legs are yellow. Males weigh about 8.5lbs (3.9kg) and females about 6lbs (2.7kg).

The Chantecler can survive in cold climates and the chicks grow well. The breed is a good layer, and a hen may produce about 200 pale-brown eggs in a year, laying throughout the winter. The bird's meat is also good, with plenty to eat on the breast and thighs. As well as its meat and egg-laying qualities, the Chantecler is an attractive, tame breed that is excellent for showing.

COCHIN

The Cochin came to the West from China in the early 1850s, and among the first to receive the bird was Queen Victoria, her interest in the breed leading to its great popularity at a time when no similar type of chicken had been seen. The Cochin is thought to have been bred by the Chinese for its feathers, used particularly to fill bed coverings. The Cochin is mainly bred as an ornamental fowl, although the hens

BREEDS OF CHICKEN

The Cochin or Cochin China, originally known as the Chinese Shanghai. Once in the US, the breed underwent considerable development to reach its current form.

make good foster mothers, and are therefore useful for rearing gamebird chicks that have been orphaned or abandoned.

The Cochin looks like a big, fluffy ball of feathers and is very rounded in appearance. It is one the largest of the heavy breeds, with males weighing about 11lbs (5kg) and females 8.5lbs (3.9kg). It is also broad in stature, which is enhanced by the huge number of soft feathers on the body. The comb and wattles are red, and are bigger in the

male, while the ear lobes are also red. The legs are yellow.

Unfortunately, Cochins are subject to metabolic and cardiac problems – partly caused by their round body shape and somewhat sedentary lifestyles. It is best to keep them on short grass, since longer vegetation can damage the feet and feathers. Wet conditions are also to be avoided, since the feathers of the legs and feet may suffer and mud balls can form on the soles of the feet. The birds can be kept in fairly confined spaces and do not fly, so a fence about 2-ft (0.6-m) high should be sufficient to retain them.

Cochin hens are excellent brooders with calm, maternal instincts. Eggs are large, although not produced in great

OPPOSITE, BELOW & PAGE 174 LEFT: Cochins come in white, buff, black, blue, cuckoo and partridge, although variations continue to be introduced. Until new colours are officially approved, such birds must be exhibited in non-standard classes.

numbers, and the chicks are strong when hatched. They mature within two years and survive until about the age of eight or ten. Cochins are docile and friendly, and being submissive can be successfully kept with other chickens, attributes which also make them good pets. They need good-quality feed to stay in top condition. Several accepted varieties are available, such as black, blue, buff, cuckoo, white and partridge.

DOMINIQUE

The Dominique is considered to be America's oldest chicken breed. It was first developed from stock that gave rise to southern English breeds, such as the Sussex and the Dorking, and which was

OPPOSITE: The Dominique is a very old American breed. It is a hardy chicken and produces excellent, dark-brown eggs (see above).

CHICKENS

brought to New England during the time of the first colonists. The breed had grown in popularity by the 19th century, and was to be found in many parts of the USA. The breed's popularity, however, then declined steadily – mainly because the Plymouth Rock breed derived from the Dominique found favour instead – and by the 1950s it was almost at the point of extinction; the American Livestock Breeds Conservancy listed it as Critical in the 1970s. But an upsurge of interest in rare breeds helped to revive the Dominique's fortunes, and today its status seems more secure.

Dominiques are a dual-purpose breed, prized both for their meat and for their brown eggs. They weigh 6–8lbs (2.7–3.6kg) when fully grown. At one time their feathers were much sought after as a stuffing material for pillows and mattresses. The birds have a distinctive, handsome appearance, with rose combs and a thick plumage of barred or

The Dominique has distinctive barred feathers.

irregularly striped black-and-white feathers. This pattern is also sometimes called 'hawk colouring', and apart from providing good insulation for the birds, also helps conceal them from predators. The breed is quick to mature, and eggs are produced at about six months of age. Dominiques are calm and steady birds – traits which make them good show birds and pets. The hens make excellent mothers, and have a high rate of success rearing chicks. They lay light or dark-brown eggs and show a lesser tendency towards broodiness than some other exhibition breeds. The meat is also of good quality.

The Dominque is hardy and is also a good forager, a trait which served it well during its early development as a breed. As well as the standard bird, a bantam variety is also available.

DORKING

One of the oldest breeds of chicken, the history of the Dorking is thought to go back to the days of the Roman Empire, where it originated in Italy at the time of Julius Caesar. It is mentioned by the Roman writer, Columella, who describes it as 'square-framed, large and broad-breasted, with a big head and an upright comb'. The breed was then introduced

into Britain where much of its development took place, and where it was used to produce the Faverolles and Sussex breeds. The Dorking is believed to have been first exhibited at a poultry show in 1845, and is still kept as a show bird today.

The Dorking is a largish breed, with males of the larger varieties (silver-grey and coloured) weighing up to about 9lbs (4.1kg) and hens achieving about 7lbs (3.2kg). In the smaller white variety, males grow to 7.5lbs (3.4kg), the females achieving 6lbs (2.8kg) in weight. Bodies are rectangular, set on short legs bearing five toes. The combs are large, which means they can be prone to frostbite even though the birds are fairly hardy otherwise. Although docile, the Dorking likes to forage and needs plenty of space to ensure it stays vigorous. It is extremely popular on account of its white flesh and good flavour, but the bird is also kept for its white eggs. Dorkings may take several years to reach maturity. Hens go broody before laying.

OPPOSITE & RIGHT: The Dorking comes in three recognized varieties: white, silver-grey, cuckoo rosecomb (rare), and coloured.

FAVEROLLES

The Faverolles originated in the village of Faverolles in France, created by crossing breeds such as Dorkings, Houdans and Asiatics. The breed was first described in 1893, with the salmon variety, now the most common, appearing in 1985. Faverolles were initially bred for their meat, but they are good egg-layers and are now seen as a dual-purpose breed.

The body of the Faverolles is broad and square and the wings are small. The broad, round head bears a single upright comb, a prominent beard, and muffling. The eyes are reddish-bay. The pinkish legs are sparsely covered, with the feathering concentrated on the outer toe. The feet have five toes. Males weigh about 11lbs (5kg).

The Faverolles is a docile, agreeable bird that can be tamed easily. They often become quite affectionate and are good as children's pets, living for around six or seven years. Faverolles are also alert and active. They prefer to live in a run, but because they are not good fliers fencing doesn't need to be too high. Because of

LEFT, OPPOSITE & PAGE 182:
Faverolles come in white, black, buff,
cuckoo, ermine, and blue-laced, as well
as in the more commonly seen salmon.

their dense feathering and small combs they are quite resistant to cold weather, and can cope more readily with damp conditions than some of the other highly feathered breeds, although they may suffer from scaly leg mite. Hens are broody and make good mothers, laying medium-sized cream or tinted eggs throughout the winter. The chicks grow quickly and will develop fast if given good-quality food.

FRIZZLE

Although the Frizzle is thought to have originated in South-East Asia about 300 years ago, there are reports of the breed having been seen in Europe in the 1600s. The name Frizzle comes from the type's unusual feathering, in which a genetic modification causes each of the feathers to curl toward the head instead of the tail. Frizzles are popular as exhibition birds, particularly the smaller bantam forms, although the breed is considered rare. In fact, it was virtually extinct until its fortunes were revived by a group of enthusiasts, which initiated a breeding programme. Although the Frizzle is a recognized breed in many countries,

OPPOSITE: Being quite rare, Frizzles are mainly kept as show birds.

BREEDS OF CHICKEN

frizzling is also present in other chicken populations and for this reason is considered by some authorities to be yet another variation that can occur in any breed.

Each of the Frizzle's feathers is moderately long and of a ragged appearance, making the bird's feathers seem permanently ruffled up. Frizzles hold their bodies and their tails erect. They have short, broad bodies, rounded full breasts, long wings, and large upright tails, their legs being without feathering. The eyes, single comb, and wattles are red. Male birds weigh about 8lbs (3.6kg), with females coming in at about 6lbs (2.7kg).

Despite their unusual appearance, Frizzles are hardy birds with equable temperaments. They are suitable for keeping in outdoor enclosures or can even be allowed to range free. The hens lay brown eggs, and the chicks, once hatched, begin to grow rapidly. They

appear to have normal feathering at first, but soon begin to assume their characteristic ruffled appearance.

The Frizzle is available in a variety of colours, including black, white, buff, blue, black-and-red, and cuckoo, with the colour of the beak corresponding to the colour of the feathers. Three types of feathering can be present in the breed: frizzled, overfrizzled and flat-coated.

LEFT & OPPOSITE: Frizzles, besides being primarily exhibition birds, included in the American Standard of Perfection, are also good egg-layers and make excellent table birds.

ABOVE: A Leghorn chicken with frizzled feathers, demonstrating that the characteristic can occur in other breeds.

CHICKENS

HAMBURG

Despite its German-sounding name, the Hamburg is an old breed that originated in The Netherlands, the breed shape having been further refined by English fanciers over a century ago. Known as Dutch Everyday Layers, their other name, Moonies, derive from the crescent moon-shaped spangles in the feather pattern. The Hamburg is an attractive breed which is kept mainly as an ornamental bird, being available in several varieties. It lays small white eggs. The birds are neat in appearance, with pea combs. Males weigh in at 5lbs (2.3kg), the females achieving 4lbs (1.8kg).

Hamburgs are quite alert and flighty chickens, capable of flying reasonably long distances. They are hardy and good at foraging, but they do not take kindly to confinement. Hamburgs are not especially docile and do not go broody.

The Hamburg comes in varieties such as golden-spangled, silver-spangled, black, white, golden-pencilled and silver-

OPPOSITE: Hamburgs are trim and stylish birds, with delicate features and a wild nature. Although good layers, their eggs are often very small.

RIGHT: Japanese Bantam chicks.

pencilled, some of which were developed in England and some in The Netherlands.

JAPANESE BANTAM

Also known in many parts of the world as the Chabo, thought to originate from a Javanese word meaning 'dwarf', evidence suggests that the Japanese

Bantam originated in South-East Asia, where it is still popular today.

This little chicken graced the gardens of the Japanese aristocracy for well over 350 years. The bantams began to appear in Japanese art in the mid-17th century, at a time when Japan was closing its doors to the outside world. It also appears in

OPPOSITE: A white Japanese Bantam, its flamboyant black tail extending well above its head.

RIGHT: Originating in China, Langshans were imported to North America in 1878 and were admitted to the standard in 1883, with white Langshans achieving this recognition later in 1893. There are three varieties of Langshans that have been accepted to the US standard; black, white, and blue, although the latter was not accepted to the standard until 1987.

Dutch art of the same period, suggesting that Dutch traders possibly carried Chabos as gifts to the Japanese from spice ports in Java, which the Dutch ruled at that time.

This is a bantam breed, the birds having large, upright tails that can extend higher than their heads, while the wings angle down and to the back along the sides. The birds are friendly and make good pets, being fond of riding on the shoulders of their owners. They are capable of fending for themselves if provided with a large enough area in which to forage, but beware of putting too many cockerels together; this may lead to fights despite their well-known friendliness. These chickens have been

known to live for up to 13 years with proper care.

Japanese Bantams come in the following varieties: black-tailed white, white, buff, black-tailed buff, grey, blue, barred, black-breasted red, black, and many more.

LANGSHAN

The Black Langshan breed originated in China. Some were imported into Britain by the late Major Croad, and a strain known as the Croad Langshan was subsequently produced. The Langshan of today is a German breed developed from

the Croad Langshan. Langshans became popular in the United States in the later part of the 19th century, having first arrived there in 1878. These days they are kept mainly as ornamental fowl, but sometimes also as a dual-purpose breed, although the narrow legs and long body are not ideal for today's meat markets.

Langshans are large, tall birds with long legs and tails that are carried high, giving them their characteristic stance. The tail is long and well-spread, and may measure 17in (43cm) in length. Their relatively small heads bear single red combs and red wattles. Males come in at a weight of 9.5lbs (4.3kg) and females 7.5lbs (3.4kg).

While Langshans are active, fast-moving birds, they are also friendly and can easily be tamed. Hens lay brown eggs and go broody and make good mothers. Varieties available include blue, white and black, the black variety showing a dark-green iridescence in the feathers.

LEGHORN

Another breed of chicken that takes its name from its place of origin, namely the port of Leghorn (Livorno) in Italy. It was subsequently further developed in other countries in Europe as well as in the USA, producing, from the original white

OPPOSITE: The German Langshan was created by crossing Croad Langshans with Minorcas and Plymouth Rocks.

RIGHT: Leghorns are reliable layers of around 300 white eggs per year.

form, the many colour varieties that are seen today.

The Leghorn weighs about 3–4lbs (1.4–1.8kg). Male Leghorns have especially prominent red combs, mostly single but sometimes double, and large wattles. Ear lobes are white, beaks are short and stout, the long legs are yellow, and in all colour varieties the eyes are red. Varieties include black, white, brown, blue, cuckoo, mottled, partridge, buff and pyle.

Leghorns are often used as laboratory animals for research, and the breed is one of several used to produce modern battery hybrid egg-layers. They are prolific and productive layers throughout the year that can adapt to different types of conditions, although the large combs are prone to frostbite. The hens seldom go broody and are non-sitters, laying good-sized white eggs; the chicks are easy to rear, and grow quickly. Leghorns are happy to roam freely, roosting in trees if given the chance, but

THIS PAGE & OPPOSITE: While they can adapt to life in a chicken run, Leghorns prefer to run free, roosting in trees whenever possible.

they can be kept equally successfully in a run. They are somewhat excitable, garrulous and active, becoming quite tame while preferring not to be handled. This breed is the foremost large-scale commercial egg-producer in the USA.

MALAY

This is an old breed that has changed very little over time. It is not known why it is called the Malay, for there is evidence that its descendants came from north India. The Malay is somewhat different in appearance from many other chicken breeds. First, it has a slim, well-muscled body and an upright stance, making it appear taller and rather more 'wild' than some of the plumper, stockier breeds often seen. It also has very long legs. The lean look is the more pronounced because the birds are close-feathered, and often carry their bodies with their long necks held erect and their tails slightly drooping. The skull is large, and the bird's protruding beetle eyebrows give it a slightly fierce expression. Males grow to 9lbs (4.1kg) in weight, with females reaching 7lbs (3.2kg). The Malay was introduced to Britain in about 1830, and the black-

ABOVE RIGHT: The Malay is an ancient Asian breed with a rangier, wilder look than other domestic chickens.

OPPOSITE & PAGES 196–197: While Marans can occur in a black variety, by far the most appealing are the cuckoos, with their dark-grey or silver barred plumage.

breasted variety was admitted to the standard in 1883. Other varieties (see below) came much later.

In Asia, this chicken is seen around villages, but it is not especially prolific as an egg-layer (young hens may lay up to 100 or so brown eggs per year, but this diminishes to about 40 eggs in older hens). Another problem is that the bird's

legs are so long that it is unable to fit easily into a standard nest.

The Malay's meat content is not very high, either, and the Malay is kept in the West chiefly as an ornamental bird or for showing at exhibitions, although this is not common. Malays are strong, hard-feathered birds that need plenty of opportunities to exercise in order to stay

in top condition. Malay blood is present in many of the best-known early breeds, as it was often used to inject vigour into stock when creating new types. Most Malays are red with black breasts, but other varieties, such as black, white, red pyle and spangled, are also seen. There is also a bantam form.

MARANS

The Marans is a breed of chicken that was developed in France, around the town of Marans, near La Rochelle. It is thought to contain elements of breeds such as Faverolles, Croad Langshan and

Barred Plymouth Rock in its makeup. It reached Britain in the 1920s, where its chocolate-brown eggs soon made the breed popular.

Marans have broad, deep bodies and an overall attractive appearance. The large single red comb may have up to seven serrations and the ear lobes and wattles are also red. The beak is medium-sized and light in colour, and the eyes are orange-red. The medium-length legs are light in colour and unfeathered. The most popular colour – and perhaps the most appealing – is the cuckoo, ranging from dark grey to silver. A black form also exists, which has a beetle-green sheen to the feathers. Males weigh 8lbs (3.6kg), whereas females weigh 7lbs (3.2kg).

The Marans is an active and hardy chicken, popular in gardens where it keeps down the population of slugs, snails and other pests. Marans do not make especially good pets, however, since they are not generally keen on being handled, although this temperamental tendency varies a little from strain to strain. A discussion with the breeder will probably determine the sort of characteristics of any particular strain that you may be about to purchase. The Marans is a good layer and also produces good meat for the table.

MINORCA

The Minorca was originally known as the Red-Faced Black Spanish. It is the largest and heaviest of the Mediterranean breeds, and good specimens usually have a stately appearance. The body is long and angular, making the birds appear even larger than they are. The tail is also long, and the large, wide feathers are held close to the body. The red comb is very prominent, the wattles are also long, and the ear lobes are white. Males range from 8–9lbs (3.6–4.1kg) in weight, the females achieving 6.5–7.5lbs (3–3.4 kg).

Minorcas are alert birds that can forage for themselves quite well. They produce large white eggs, and yields are quite good. They are rather poor meat producers, however, due to their narrow bodies and slow rate of growth, but they are in any case mainly bred for their exhibition qualities. Varieties include single-comb black, single-comb white, single-comb buff, rose-comb white and rose-comb black.

NEW HAMPSHIRE RED

This is another of those breeds that looks like a 'typical' chicken. The breed was developed over a number of years, beginning in about 1915, from specialized selections of the Rhode Island

Red brought into the American state of New Hampshire. Since then, the breed is not known to have been diluted with any outside blood. The New Hampshire Red was admitted to the standard in 1935. Today its best qualities – and the ones sought after by breeders – are fast maturation, quick feathering, the production of large brown eggs, and general strength and vigour. Bred initially for its eggs, the bird is now a dual-purpose breed and its meat is used either for broiling or for roasting. Special strains have been developed that exhibit rapid growth and the accumulation of increased weight, but which are lacking the egg-laying qualities of the standard dual-purpose strains.

The body is deep-breasted and well-rounded with a medium-length tail. The head is deep and fairly flat on top with prominent eyes. There is a single comb with five points (that may fall slightly to one side in females), large wattles, and

OPPOSITE: The Minorca is a moderate layer, but is prized for its show qualities.

RIGHT & PAGE 200: The New Hampshire was initially used in the Chicken of Tomorrow contests, a precursor of the modern broiler industry.

red ear lobes. The legs are yellow, the lower thighs being large and muscular. The feet have four toes. The feathers are a rich chestnut red that have a tendency to fade in strong sunlight. Males weigh 8.5lbs (3.9kg) and females 6.5lbs (3kg).

New Hampshire Reds do well either in runs or when allowed to range free. They aren't good fliers, so high fencing is not needed. They can cope with cold weather, but the combs are prone to frostbite, so care should be taken in such conditions. Hens lay well, even in winter; they are prone to broodiness and make good mothers.

ORPINGTON

Coming from the town of Orpington in the English county of Kent in the 1880s, this breed was introduced to the USA in the 1890s, where it quickly achieved popularity due to the quality of its meat. The first Orpington, a black, was developed in 1886 by William Cook, who used Langshans, Minorcas and Plymouth Rocks to create the breed. He then went

OPPOSITE & PAGES 202 & 203: The Orpington's large size and soft appearance, together with its rich colour and gentle contours, make it one of the most attractive of the breeds.

on to develop the other solid-coloured varieties. The Orpington is a dual-purpose chicken reared for its eggs as well as its meat. It is also popular as a show bird.

Because of its heavy, loose feathering, the breed can appear massive and stocky. However, this generous feathering also allows it to withstand colder temperatures than many other breeds. The feathering is extensive, and almost covers the legs. The birds have an upright stance, and heads that appear small in relation to the size of the body. Male Orpingtons weigh 10lbs (4.5kg) and females 8lbs (3.6kg). The prominent combs are red, as are the wattles and cheeks. Dark-coloured varieties (see below) have dark eyes and legs, whereas

pale-coloured varieties have red eyes and
white legs.

Orpingtons adapt well to either free-
range living or being confined in runs.
They are fairly placid and tame, and
when kept together seem to be non-
aggressive toward other chicken breeds.
Hens show broodiness, laying small

new colours are also being bred, such as porcelain and red.

PEKIN

The Pekin is said to have been taken from the private collection of the emperor of China in Peking (Beijing) by British forces in 1860, while another story describes how they were exported from China even earlier and presented to Queen Victoria. Either way, the Pekin, which is a true bantam variety, is known throughout most of the world as the Miniature Cochin, which can lead to confusion since the Pekin is a distinct breed and is not really a miniature version of a Cochin at all. No doubt some of this confusion arose because 'Cochin' was once the term used for all birds of this type coming from China, although today's Pekin bears little resemblance to the birds brought from that country all those years ago; these were taller and generally bigger, holding themselves in a more upright stance.

Beneath the fluffy exterior, the Pekin has a short, robust body, the head being set slightly lower than the top of the tail. This gives the impression of the bird bring tilted slightly forward – a feature regarded as important in showing circles. The small head, bearing a serrated comb,

brownish eggs (about 150 per year), and make good mothers. Orpingtons like to exercise, but they have short wings and can therefore be kept in enclosures with low fences.

Orpingtons exist in various solid-colour varieties: black (with a single comb or rose comb), blue, buff and white. The lighter coloured buffs and whites are smaller than the darker varieties. Some

CHICKENS

BELOW: The Pekin is a bantam variety originating in China. It has a fluffy appearance and its body is stocky and robust.

is set on a short neck, surrounded by a hackle that reaches well down the back.

The Pekin's legs are short and almost invisible beneath the fluffy covering of feathers. The males weigh 24oz (680g), the females 20oz (570g). The Pekin comes in a variety of colours – too many according to some breed experts, who regard this overabundance as damaging to the breed standard. Colours include lavender, blue, black, buff, barred, Columbian, cuckoo, mottled, partridge and white.

Pekins have endearing personalities; they are docile and friendly and look most attractive. They make good pets for

children, although the feathering on the feet means that care must be taken to avoid them becoming wet and clogged with mud. Because they are small, Pekins can be kept in more limited spaces than might be possible for many other breeds, although they do like to forage for themselves on grass. The hens go broody and make excellent mothers, laying a fair number of tinted eggs.

PLYMOUTH ROCK
This heavy breed, which takes its name from the New England town of Plymouth, was first exhibited in Boston in 1829, at America's first poultry show. However, it is thought that the breed line of the original birds was lost, and the breed as we know it today was then exhibited again in 1869, at Worcester,

ABOVE & OPPOSITE: The Plymouth Rock was created from a variety of breeds.

OPPOSITE & LEFT: The Barred Plymouth Rock was one of the first to appear, with other varieties coming later. This plumage pattern, however, remains the most popular of all.

Massachusetts. The breed was derived from crosses between several other breeds, including Dominique, Cochin and Java, and possibly Dorking and Malay. The first Plymouth Rocks had barred plumage, and were in fact called Barred Plymouth Rocks, although other plumage patterns were later developed. The Plymouth Rock gained full breed recognition in 1874, when it was included in the American Standard of Excellence.

The Plymouth Rock became one of the foundations of the broiler industry of the 1920s, and was the most widely kept and bred chicken in America before the Second World War. This was on account of the breed's all-round qualities, such as its hardiness, docile nature, excellent meat, and its egg-laying abilities. The eggs are brown, but the particular shade may vary considerably. A hen may lay about 200 eggs in a year, and since the breed is prone to broodiness, frequent egg collection is advisable. Plymouth Rocks are big chickens and the hens

have deep, full abdomens – an indication of a good egg layer. The breed is also characterized by having a broad, deep and well-rounded breast and bright yellow legs. The wattle, ear lobes and comb are red, and the bird has a bright-yellow beak and bay- coloured eyes. Males weigh 9.5lbs (4.3kg), with females coming in at 7.5lbs (3.4kg).

Plymouth Rocks are friendly birds that are easy to tame. Although they prefer to run free they do not require a great deal of space, and since they are poor fliers, fencing doesn't need to be built especially high. The chicks feather up quickly and the breed makes a good pet for children, being of a calm temperament as well as long-lived.

The breed comes in several varieties, including barred, buff (first exhibited as golden-buff), Columbian, partridge, silver-pencilled and white.

POLAND (POLISH)

The Poland is an ancient breed whose origins – and even the reason for its name – are unclear. Crested birds of this type have been described in many parts of Europe, not only in Poland, and it was known as a pure breed as far back as the 16th century. It was shown at the first poultry show in London, England, in 1845, where it was able to achieve breed classification.

As mentioned, this is one of the crested breeds. The bird has an upright stance, a tightly feathered body, and showy, erect tail feathers, but it is the head that attracts all the attention – the crest seemingly having a central parting, with the feathers fanning out on either side. Some also have a beard and muffs. Breeders and keepers of this chicken sometimes tie the crest up to keep it clean and to enable the bird to get a better view of the world. The crest of the male is spiky, whereas that of the female is more rounded. The eyes are red in all colour varieties. Males grow to about 6lbs

RIGHT & OPPOSITE: The heads of Polands are adorned with large crests, caused by a cone or protuberance on the top of the skull. The crest occupies almost the entire head.

OPPOSITE & THIS PAGE: Prone to hypothermia, the Poland needs to be kept in a warm environment during cold weather.

(2.7kg) in weight, with females achieving around 4.5lbs (2kg).

The Poland is described as a non-sitter, laying white eggs, but it can occasionally become broody. It is best to use an incubator. It needs plenty of space in which to roam, as overcrowding can lead to crest-plucking behaviour. It

should also be kept warm in cold weather, because it has a relatively thin skull and can quickly suffer the effects of hypothermia. Furthermore, ice can form in its crest when it drinks, which can cause problems, as can mites, which the bird finds difficult to remove by preening.

The Poland is available in various colours, the best-known variety being the white-crested black, i.e. a black body with a white crest. The other two similarly-patterned types are the white-crested blue and the white-crested cuckoo. In these aforementioned colour varieties there is no beard but the birds do have wattles, while in all other colours the birds have beards but no wattles. The ear lobes are white. Beaks and legs are dark blue in all the colour varieties apart from the white-crested cuckoos, which have paler beaks and legs. Chamois, gold and silver birds have laced feathering.

The Poland comes in several colour variations, including chamois (opposite) and gold (right).

BREEDS OF CHICKEN

RHODE ISLAND RED

Developed in the eastern American state of Rhode Island around the 1890s, the breed occurred by way of crossings with other breeds, including Leghorns, Malays and Cochins. It is from the Malay that the Rhode Island Red derives its deep-red feathers. Original birds had both single and rose combs, or even pea combs. The Rhode Island Red is an important dual-purpose bird, and today it is arguably the best-known breed throughout the world.

It was introduced to Britain in 1903, where its popularity expanded quickly, and crossings of the Rhode Island with the native Sussex breed produced many of Britain's modern chicken hybrids.

As the name suggests, the Rhode Island Red has rich, dark, glossy red plumage, this being especially handsome in the males. Black feathering can be present in the wings and tail, but only the female should have black feathers on the neck. The comb, wattles, eyes and ear lobes are also red, and the legs are yellow. The body is rectangular, broad, and deep, with a flat back and a medium-sized tail. Males weigh in at about 8.5lbs (3.9kg) and females at 6.5lbs (3kg).

Although the Rhode Island Red is classed as a heavy breed, it is active and fairly hardy. It can make a good pet, but some males may be aggressive. It enjoys foraging and can cope with marginal conditions better than many other breeds. Rhode Island Reds are probably the best egg-layers of the dual-purpose breeds, producing plenty of brown, or dark brown, eggs.

OPPOSITE, RIGHT & PAGES 222 & 223: The Rhode Island Red is probably one of the best-known breeds worldwide.

BREEDS OF CHICKEN

SCOTS DUMPY

The Scots Dumpy is thought to have originated in Scotland, the 'dumpy' part of the name, meaning short and stout, referring to the breed's thick-set appearance and extremely short legs. But historically they have been known under a host of names, such as Bakies, Stumpies, Dadlies, Hoodies, and Creepies, and were probably the birds referred to in Gaelic as *Coileachchime* and *Coileach degh sheinneadair*.

Shortness of leg is the breed's most defining characteristic, producing a waddling gait, with adult birds being less that 2in (5cm) off the ground. The birds

Although the exact origins of the Scots Dumpy are unclear, evidence suggests that birds with similar physical characteristics were present in Britain several hundred years ago.

are also longer in the back and have tails that are set lower than in other breeds.

There is no set colour for the Scots Dumpy, but cuckoo, black, and white are the most common. Other patterns/colours occur as sports, or have been bred into the phenotype by fanciers. The breed's standard allows for any pattern/colour that is allowed in game fowl, although colour is only worth 10 per cent of the score in competition.

On the whole, this is a docile breed but, as in any breed, occasional males can be aggressive. Hens are good layers of light or white to tinted eggs, and are said to make good table birds.

SEBRIGHT

Another true bantam with no standard form, the Sebright is one of the most popular of breeds, due largely to the extraordinary laced appearance of its feathering. It bears the name of Sir John Sebright, the person who developed it over 200 years ago, and although there is some doubt concerning the actual breeds used in his endeavour, it is likely that they included Nankin, Poland and Hamburg chickens. Sebright established a club for his breed in 1810, and

The Sebright is a true bantam, in that it is a miniature bird with no corresponding larger fowl to which it is related.

Sebrights come in gold and silver varieties, both with the distinctive lacing on the feathers.

development continued, including the creation of the gold and silver varieties seen today. The breed standard was established in 1952.

Although small, Sebrights carry themselves with confidence. The body is compact, having a short back and a full, puffed-out breast. The wings seem slightly large, given the size of the bird, and they are carried low and angled down toward the ground. The tail is well-fanned and carried high. Male Sebrights usually lack the big, curved sickle feathering that is seen in the tails of most males, having what is known as hen feathering instead. The result is that the feathering appears sleek and neat in both sexes.

The head is small with a short beak which is dark blue or horn-coloured in silver varieties, but a darker horn-colour in golden varieties. Males have rose combs with fine points. The feathering on the gold variety appears as a golden-bay background coupled with the strong, well-defined black lacing that has a greenish sheen in the light. The feathers of the silver variety have a silvery-white background colour with the same black lacing. In both varieties the legs and feet are slate grey. Male Sebrights weigh 22oz (620g), the females coming in at 18oz (510g).

The Sebright is a difficult bird to rear, with a high mortality rate in its first few weeks of life, making it unsuitable for novice breeders. The attractive feathering makes the bird a favourite with exhibitors, but it is also a suitable breed to keep as a pet in the garden, even as a member of a mixed flock. Sebrights are curious and good-natured, being hardy and with a fondness for ranging free. They also like to fly. Eggs have light-coloured shells, although egg-laying performance varies and the hens are not especially broody.

CHICKENS

SILKIE (SILKY)

A most unusual-looking chicken, the Silkie is a small variety that ranks among the 15 or so most popular breeds in the USA. Of ancient Asian stock, the Silkie's exact origins are shrouded in mystery, as is the case with many of the older breeds. What is not in doubt, however, is that the

The first written account of what may be the Silkie breed comes from Marco Polo, who saw chickens with fur-like plumage during his 13th-century travels in China.

Silkie has been known for hundreds of years. The great traveller, Marco Polo, describes seeing a 'furry chicken' on his

travels in China in the 13th century, and the description is in some ways accurate, for the unusual feathering of the Silkie does resemble a covering of soft fur. When trade routes were established between East and West, the Silkie was brought to Europe; much later, the bird was advertised and sold in The Netherlands as a cross between a chicken and a rabbit!

As mentioned, the Silkie has fur-like feathering, the unexpectedness of it producing an almost comical appearance, and it is this soft-looking plumage that has given the bird its name. The look is completed by a powder-puff crest on the top of the head and in some varieties even fluffy ear muffs and a beard. The feathering extends right down to the feet. The skin, including the comb and wattles, is a purple-blue colour, with blue ear lobes. When produced for the table, the unusual dark skin seems very strange in a cooked bird – a surprise heightened by the fact that the bones are also nearly black in colour.

Silkies are small birds, with males weighing 4lbs (1.8kg) and females 3lbs (1.4kg). Although this small size sometimes confuses people into calling them bantams, they are in fact a small example of a standard breed. A true

OPPOSITE, RIGHT & PAGE 232: Silkies are calm, gentle birds that prefer to be with their own kind, making them unsuitable for mixed flocks.

bantam variety also exists, however, which is only two-thirds the size of the standard bird. It was officially recognized in Britain in 1993.

Male Silkies have broad, full bodies with short backs. Tails and wings have a ragged appearance, especially at the tips of the feathers. The head bears a walnut-type comb, black eyes, and a short beak. The legs are dark grey. This is another breed in which the foot has five toes. Females have shorter legs, with their underfeathers almost touching the ground. The comb, wattles and ear lobes are smaller in the female. The Silkie comes in black, white, blue, gold and partridge colours.

Silkies are calm and gentle birds and could almost be termed the perfect pet but for their need to avoid wet or muddy conditions, which can cause problems with the plumage and promote mite diseases such as scaly leg. Feeders and water containers should also be positioned so that feathers do not become soiled, since this can again lead to parasitic infections. Silkies do not fly and

RIGHT: The Sussex was bred as a dual-purpose bird and is one of the most productive breeds. Large eggs are laid that are cream to light brown in colour, and approximately 240 to 260 eggs can be expected in a year, the light and white varieties of hen being the best layers. Recently, some light Sussex hens produced olive-green eggs, although such occurrances are extremely rare.

don't seem to mind being reasonably confined. They don't mix well with other chickens, however, so need to be kept apart. The Silkie is a good mother but spends a lot of time being broody, so egg production will not be prolific; about 100 eggs per year may be expected.

SUSSEX

This attractive breed comes from the southern English county of Sussex. One of the oldest chicken breeds, the Sussex was a popular bird over 100 years ago and is still so today. Although bred as a dual-purpose bird, the Sussex chicken is a prolific egg-layer. It is now found in many countries, where it is often a typical 'backyard chicken'. The breed is also exhibited at shows.

The Sussex is a graceful, well-proportioned bird with a flattish back and

LEFT: The bantam version of the Sussex chicken.

BELOW: Speckled hens are more prone to broodiness than other varieties of Sussex chicken.

with the tail held up at an angle of about 45 degrees from the body. Feathering is quite dense and also extends well down the legs. The medium-sized red comb is single and erect, and the wattle is fairly long in the male bird. The ear lobes are also red, as are the eyes in the dark-coloured varieties, being orange in the lighter colours. The legs are white. Males weigh about 9lbs (4.1kg) and females about 7lbs (3.2kg).

The Sussex is a docile bird and a good forager, and adapts well to a free-ranging life, although it can also be kept successfully in more confined conditions. Hens may sometimes go broody, especially those of the speckled variety.

OPPOSITE & RIGHT: A light variety of Sussex cockerel.

About 250 large, cream or light-brown eggs are produced each year, and resulting chicks are fast to mature.

Different varieties include:

White This is pure white.

Speckled The feathers have a mixture of dark brown and black, with white tips.

Brown The cock is dark brown with black points, whereas the hen is paler.

Light A white body with black wingtips and tail. The neck is black-and-white.

Coronation This has similar colours to the light, but the black markings are replaced with blue-grey.

There is also a bantam form, which may be seen in any of the varieties appearing in the standard bird.

WELSUMMER

The Welsummer gets its name from the Dutch village of Welsum, despite the fact that it was actually developed in the region to the north of Deventer, at about the same time (1900–13) that the Barnevelder breed was being created. The Welsummer arose from crossings between the Cochin, Wyandotte, Leghorn, Barnevelder and Rhode Island Red. In 1921 the

CHICKENS

Welsummer was exhibited at the World Poultry Congress in The Hague, Holland, and in 1928 was introduced to Britain. It is a fairly common breed today.

The Welsummer male has the look of a typical farmyard cockerel, being upright

LEFT, BELOW & PAGE 240: The Welsummer was developed a little after the turn of the 20th century, its main characteristic being its large eggs, described as being 'a rich, deep flower-pot red'.

and with a broad back and big, full tail held erect. The head bears a large red comb, medium-length wattles, almond-shaped ear lobes, and a short, strong beak. The legs are yellow. Males weigh about 7lbs (3.2kg, the females 6lbs (2.7kg).

Welsummers forage well when allowed to range free, although they can also be kept in runs. Hens go broody and lay large, brown, pigmented eggs, although there are fewer in winter. Chicks are easy to sex because males have lighter-coloured heads than the females, and back markings. Varieties such as silver duckwing, gold, and black-red partridge are possible.

These are friendly, easy-to-handle birds which live for about nine years.

WYANDOTTE

The Wyandotte originated in the United States, although the breeds used in its creation are not clear. The first variety, the silver-laced, was developed in New York State in the 1860s, where it became known as the American Sebright or Sebright Cochin. The other varieties were created in the north and north-eastern states in the 19th and 20th centuries, and the Wyandotte is now a fairly common breed.

Wyandottes are large chickens with a noticeably rounded appearance, their

bodies being broad and full-feathered. Hens are deep-breasted, indicating they are good egg-layers. The Wyandotte has a short, round head, a rose comb, bright-red ear lobes, and reddish eyes. The legs are yellow. Males weigh about 8.5lbs (3.9kg) and females 6.5lbs (3 kg).

Wyandottes are a good dual-purpose breed. They tend to be docile birds, and the hens make good mothers,

THIS PAGE & PAGES 242 & 243: The Wyandotte is an attractive bird with unusually beautiful plumage. The silver-laced variety is pictured here.

CHICKENS

being productive layers of brown eggs, and have strong, quick-growing chicks. They have good temperaments, an attractive body-shape and plumage-patterning, and an ability to remain healthy even in adverse conditions, making them popular with chicken fanciers and farmers.

The Wyandotte is available in several colours and patterns, including white, blue, buff, black, partridge, silver-pencilled, silver, gold, blue, buff-laced and Columbian – which is not so very different from the plumage of the light Sussex.

GLOSSARY

Like most areas of special interest, the world of chickens has its own language. Here, the most commonly encountered terms are explained:

Bantam: A small version of a domestic chicken, about one-quarter the weight of a standard bird. Some breeds are available in bantam form and some are not. A few breeds are only available as bantams.

Barring: A form of feather patterning in which equal-sized, alternating stripes run across the feather.

Beetle brow: This is a form of feathering that gives distinct, heavy eyebrows, such as seen in Malays.

Breed: In this context, a breed is any individual and distinct type of chicken, such as a Cochin, a Leghorn, or a Rhode Island Red.

Broiler: This is a young chicken used for its meat that can be cooked tender by broiling (grilling): a method of cooking using direct heat. Broilers usually weigh between 2.5–3.5lbs (1–1.6kg).

Brooder: A heated system used for artificially rearing chicks.

Broody: Describes a hen's condition in the period during which she is intent on incubating her eggs and is often reluctant to move off the nest until they are hatched.

Cape: Describes the feathers between the neck and shoulders.

Capon: A capon is a cockerel that has had its reproductive organs removed, usually at 6–20 weeks of age. Caponization produces less aggressive birds that may be used to 'mother' chicks. Capons produce plenty of tender meat, which because of its high fat content is self-basting. The practice of castration is now illegal in some countries.

Chick: This is a baby chicken, although the term is also used to describe the young of any bird species.

Cockerel: A young, uncastrated male chicken or rooster.

Comb: The fleshy structure on top of a chicken's head. It may be variously shaped, and is usually bigger in the male.

Coop: This is a secure, dry, purpose-made structure in which chickens are put during the night, to keep them safe from predators and the worst of the weather. Also sometimes called a roost.

Crest: The feather arrangement on a bird's head.

Cross-breeding: The process of mating together two different breeds or varieties to produce an individual with enhanced characteristics.

Debeak: A practice by some commercial chicken enterprises whereby the tip of the chicken's upper mandible is removed to prevent the hens from pecking each other.

Dual-purpose chicken: This is one that is kept, or valued, both for its eggs and for its meat.

Dust-bath: A receptacle containing clean earth, sand or wood ash in which a bird can 'bathe' to help rid itself of parasites and generally keep the feathers in good condition.

Ear lobe: A patch of skin on the side of a chicken's head, below the ears.

Free range: Describes the lifestyle of a chicken that is allowed to roam freely

over a reasonably large area of open ground, such as a smallholding, foraging for its own food as well as usually being provided with food by its keeper. The accepted maximum density for free-range birds should be 50 birds per acre (0.4 hectares). Today the term is often misused to give the impression that chickens are reared in 'ideal' conditions.

Frizzled: Describes a feather that has a pronounced curl, usually pointing toward the bird's head.

Hackle: A type of feather growing on the neck of a chicken.

Heavy breed: A breed of chicken in which the female has a weight of more than 5.5lbs (2.5kg).

Hen: A female chicken over a year old.

Hybrid: An individual created by cross-breeding to produce specifically required traits.

Incubation: The process during which a fertilized egg develops into a chick ready for hatching. This is often accompanied by the hen sitting on the eggs to keep them warm. In other farming methods, artificial forms of heating are used instead.

Laced: This describes an attractive patterning variety seen on the plumage of some chickens, in which the feathers are edged in a different colour than the rest of the feather.

Litter: The covering material that is placed on the floor of the roost and run.

Mandibles: The upper and lower portions of a bird's beak.

Pellet: A small piece of specially prepared food containing all the required nutrients for a chicken. There are pellets designed for general growth, breeding, and so on.

Pullet: A female under a year old.

Run: An enclosure in which chickens can roam during the day. They are usually attached to, or incorporate, a roost in which the chickens can be locked safely away at night.

Rooster: An American term for a male chicken.

Saddle: A type of feather growing on the back of a chicken.

Scale: One of the many small, overlapping, keratinous structures that cover the legs and feet of birds.

Sickle feathers: The long, curved feathers in a male bird's tail.

Variety: Describes a specific variation, such as plumage colour, plumage pattern, or comb structure seen within a breed.

Vent: The aperture through which eggs and faeces are voided from the body.

Wattle: One of a pair of fleshy, usually pendulous structures, that hang from the side of the mouth of a chicken and assist heat regulation.

INDEX

ACKNOWLEGDGEMENTS

Front cover: © istockphoto/suemack
Back cover: © istockphoto/Noam Armonn
Spine: © istockphoto//Enfys

The following images were supplied by © Art Directors & TRIP Photo Library/ Brian Gibbs: pages 22, 23, 218 right, 120, 155, 156, 157, 158, 159 both, 165, 174 left, 181, 182, 185, 189, 190, 192 right, 196, 198 left, 210, 211 both, 212, 213, 223, 231, 234, 235, 236, 241 left, 242, 243. Helene Rogers: pages 64, 68, 69, 70, 71 both, 114 above, 200. Jean Hall: page 199 right. Mark Stevenson: page 229 right. Ron Deadman – deceased: page 11.

Library of Congress: page 65

The following images were supplied by © istockphoto, courtesy of the following photographers; Aaron Holbrough: page 140. Aldo Ottaviani: page 115. Alexander Wurditsch: page 164. Anasem: page 116 left. Andrew Dean: page 29 right. Andy Gehrig: pages 54, 103, 162. Anzelitti: pages 44-45. Asist: pages 90-91. AtWagG: page 119. Brandon Laufenberg: page 40 right. Cabanariver: page 184 left. Carole Gomex: pages 107, 224. Cathleen Abers-Kimball: page 216-217. Chris Putnam: page 178. Chris Schmidt: page 129. Christian Carroll: pages 84-85, 137. Christine Nichols: pages 188, 201. Cornelia Pithart: page 166. Dana Milstead: page 221 left. Dan Chippendale: page 142. Daniel Mar: page 218 left. Dave Brenner: page 122. David Kay: page 221 right. David Peskens: page 3, 161. DaydreamsGirl: page 48. Debbi Smirnoff: page 222. Denice Breaux: page 203. Dieter Hawlan: pages 58, 112-113. Dirk Freder: pages 37, 133. Eddie Green: page 55. Edd Westmacott: page 233. Edward Ralph: pages 82, 171, 197. Edyta Cholcha-Cisowska: page 149. Enfys: page 226. Eric Naud: page 128. Evrim Sen: pages 62-63. Ekaterina Starshaya: page 47. Frank van den Bergh: pages 6, 111. Garry Adams: page 209. Geoffrey Holman: pages 8-9. Georg Clerk: page 138. Gretchen Zufall: page 15. Howard Oates: pages 20-21. Iain Sarjeant: pages 80, 108. Ilker Yüksel: page 4. Irina Igumnova: page 232. Jan Kowalski: pages 104-105. Jennifer Daley: page 126. Jerry Moorman: page 175. Jim Atherton: page 30-31. Johnnyscriv: pages 100, 106, 160. Juan Nel: page 246. June Lloyd: pages 153, 248-249. Karen Stork: page 192 left. Kary Nieuwenhuis: pages 254-255. Lauri Wiberg: pages 144-145. Kati Molin: page 208. Lawrence Sawyer: pages 75, 131. Linda Alstead: page 114 below. Linda Steward: pages 238, 241 right. Lisa Christianson: page 173. Lowell Gordon: page 191. Lucinda Deitman: page 183. Ludovic Rhodes: page 215. Mark Bond: page 227. Mark Lijesen: page 39. Mark Rasmussen: pages 78-79. Mary Lee: page 169. Mary Lee Woodward: pages 139, 239, 240. Michael Westhoff: page 59. Mike Dabell: pages 24, 179. Noam Armonn: page 77. Pamspix: 130. Parpalea Catalin: pages 2, 194, 195. Paul Hart: pages 40 left, 99, 186. Peter Bradshaw: page 35. Plourn1: page 19. Rade Pavlovic: page 67. Ramon Rodreguez: pages 72-73, 124, 134. Ra-Photos: pages 56-57. Rob Broek: pages 14-15. Robin Arnold: pages 167, 205, 206, 207. Ruchos: page 109. Sabrina dei Nobili: page 170 left. Saied Shahin Kia: page 46 above. Shannon Forehand: page 172. Sharon Kaasa: page 28 left. Shirly Friedman: page 135. Simon Burgess: pages 42, 237 left. Stephen Cridland: page 66. Steve Mann: page 76. Steve McWilliam: pages 96-97. Steven van Ringelestijn: page 163. Suernack: pages 5, 49, 117, 127. Tim McCaig: page 214. Tina Lorien:page 230. Tom Brown: page 34. Tom O'Connell: page 150-151. Tony Campbell: pages 146-147. Wojtek Kryczka: pages 50-51, 228. Yulia Sapanova: page 248. Yungshu Chao: pages 52-53.

The following images were supplied by Flickr Creative Commons, courtesy of the following photographers; 177: page 116 right. Abulic Monkey: page 26 left. Allan Hack: page 88. Alice Wilkman: page 154. Alex Harries: pages 46 below, 110. Andrea de Poda: page 74. Andy Roberts: page 25. Anne Norman: page 38 left. Arne Heggestad: page 125. Boris Bartels: page 81. Brian Kelley: page 118 right Bulaclac Paruparu: page 87 below. Charles Dawley: page 245. David Boyle: page 94. Fernando de Sousa: page 180. Filip Maljkovic: page 26 right. Garry Knight: page 60. Gina Pina: pages 10, 83, 187. Guido "random" Alvarez: pages 12-13. Hans Splinter: page 121. Jay & Melissa Malouin: pages 93, 101. Jennifer Dickert: page 220. Jhm54/Jim: page 86. Just Chaos: page 184 right. Keven Law: page 33. Kevin: page 17. Linda N: pages 36, 95. Liz Lawley: page 98. Matt Steppings: page 123. Monkeywing: page 225. Nick Stenning: page 41. Rob & Stephanie Levy: pages 18, 43, 87 above. Sean Aldrich: pages 61, 118 left. Taro Taylor: page 27. Thaddeus Quintin: pages 32, 193. Thomas Kriese: pages 202, 204. Tomasz Prezechlewski: page 102. Valerie Everett: page 89. Woodleywonderworks: page 92.

Wikimedia Commons: page 219, 237 right.
Wikimedia Commons/Eadepoeltegroen: pages 168 both. Ron Proctor: pages 174 right, 176, 177.